NUCLEAR WEAPONS:

Also published by Macmillan in association with the Oxford Research Group

Scilla McLean (*editor*)
HOW NUCLEAR WEAPONS DECISIONS ARE MADE

NUCLEAR WEAPONS: WHO'S IN CHARGE?

Hugh Miall

Foreword by
Scilla McLean

M

MACMILLAN
PRESS
in association with
THE OXFORD RESEARCH GROUP

First published 1987

Published by
THE MACMILLAN PRESS LTD
Houndmills, Basingstoke, Hampshire RG21 2XS
and London
Companies and representatives
throughout the world

Photoset in Times by
CAS Typesetters, Southampton

Printed in Great Britain by
Anchor Brendon Ltd,
Tiptree, Essex

British Library Cataloguing in Publication Data
Miall, Hugh
Nuclear weapons: who's in charge?
1. Nuclear weapons
I. Title
355.8'25119   U264
ISBN 0-333-44676-3
ISBN 0-333-44677-1 Pbk

Contents

Foreword

For years, leaders from West and East have met to negotiate about arms control. Treasury ministers agree that the new generations of ultra-sophisticated weapons cost too much and we cannot afford them. Every time there is a deployment of nuclear weapons, it is met by a storm of controversy and public protest. And yet we continue to build them – new ones, cleverer ones, more expensive ones, by the minute. Why?

This book is the result of asking that simple question. In June 1982 the United Nations held its Second Special Session on Disarmament in New York, and delegations from member nations from every corner of the world met for six weeks to try to solve the problems of world armaments. After one week of the conference, the largest peaceful demonstration in US history packed the streets of New York, filling Central Park with nearly a million people who had come to express their anxiety and concern over nuclear weapons. The next day I went back down to the steel and glass UN building to talk with various delegations and found that nothing had changed. Not one nation had budged its position one inch. The conference ended in failure five weeks later.

I came back from New York determined to understand why all this effort made no difference, why the energies and concern of hundreds of thousands of citizens and diplomats from dozens of countries could do nothing to stop, or even slow down, the arms race. In a few months, with financial support from Quaker trusts, an independent research group was set up in Woodstock, Oxford, with the aim of finding out how decisions are made to develop, produce and deploy nuclear weapons in each of the nuclear nations. The Oxford Research Group, governed by a board of sponsors, carried out this study over a period of three years, making the information available to the public as it progressed, and in 1986 a first book based on the research was published. *How Nuclear Weapons Decisions Are Made* describes how new weapons are developed, and identifies the organisations involved, in the Soviet Union, the United States, Britain, France and China, as well as in the two nuclear alliances.

In the course of our research, however, a great deal more had come to light than the separate national structures and processes of decision-making, described in that first book. There emerged remarkable similarities, and remarkable differences, in the way these decisions are made in the different countries. In comparing the roles of the various people involved – scientists, strategists, bureaucrats, industrialists, politicians – some clues began to appear in answer to the original questions posed. We knew that a second book was necessary, and that it had to be written by a researcher who not only understood the detail of all the separate processes involved, but had the perspective to take on overall views, and the ability to write the story in a simple, clear compelling way. Hugh Miall was that person, and joined the Group in April 1985.

This book is a new lens for looking at the arms race. It doesn't look at it from the point of view of trying to prove one side 'right' and the other 'wrong'. Nor does it look at it from the traditional perspectives of political science. It doesn't try to cope with it quantitatively by counting the numbers of weapons, or the comparative sophistication of the hardware. Nor does it deal with strategy. The lens of this book is decisions, how they are made and the people who make them. It will help the non-expert to link his or her own impressions of nuclear weapons with the realities of how they are designed, approved, built, deployed and explained to the public. It will enable ordinary people to understand, without becoming terrified or confused, what is happening in this extraordinary world of galloping technology and apparently endless expenditure. Lastly, it will give the reader some ideas of what might be done to bring the situation under control.

Woodstock   SCILLA McLEAN

Preface

The Vietnamese Buddhist monk, Thich Nhat Hanh, tells the following story. There was a certain man, riding on a horse. He was riding very fast, galloping along. As he flew past, a bystander shouted to him: 'Where are you going?' 'I don't know', replied the man, 'ask the horse.'

Ever since the possibility of nuclear fission arose in the minds of the physicists of the 1930s, the development of nuclear weapons appears to have had a momentum of its own. In the 1940s, the scientists at Los Alamos developed the first atomic weapons because they were sure that if they did not do so, the Germans would. Scientists today are at work on new generations of nuclear weapons, impelled by similar fears. In the aerospace companies, driven by the restless competition for contracts, designers and engineers are developing new missiles and other delivery vehicles. In the military services and the defence bureaucracies, the specialised departments responsible for new weapons are pressing for their latest projects to be approved. The military establishments, using 'worst-case' assessments of threat, are pressing for the latest weapons, and the political leaders, seeing no alternative, are ordering them. Parliamentary bodies are being excluded from decisions, or endorsing decisions after they have been made. Among the public, there is fear of a nuclear war, accompanied by a pervasive sense of helplessness.

'Over all these years', as US Ambassador George Kennan has put it, 'the competition in the development of nuclear weaponry has proceeded steadily, relentlessly, without the faintest regard for all the warning voices. We have gone on piling weapon on weapon, missile on missile, new levels of destructiveness upon old ones. We have done all this helplessly, almost involuntarily, like men in a dream, like lemmings heading for the sea, like the children of Hamlin marching blithely behind their Pied Piper.' We have been like the riders of a horse without reins.

If we wish to understand what direction we are going in, it behoves us to examine the horse. For, independent of the content of decisions, the decision-making structures which have evolved for developing nuclear weapons, which in theory are under the control of the people through the constitutional agencies of each state, in fact have a dynamic and a direction of their own.

*   *   *

The first part of this book is about the nuclear decision-makers. Chapters 2 to 6 examine the nuclear scientists, who design the nuclear warheads, the defence contractors, who make the missiles, the military establishments, who are the customers for the weapons, the defence ministry officials, who administer the development programmes, and the politicians, who endorse the final decisions. Chapters 7 to 10 ask: to whom are the decision-makers accountable? They discuss the limits to public scrutiny set by secrecy, the role of parliaments and representative assemblies, the ways in which financial control is exercised and the impact of public debate on decision-making. Chapter 11 asks – who formulates arms control and disarmament positions? Finally chapter 12 presents conclusions and certain proposals for change.

It was not my intention in this book to provide a comprehensive account of the nuclear decision-making system in any country. Readers who are interested in this may refer to the Oxford Research Group's first book, *How Nuclear Weapons Decisions Are Made*, edited by Scilla McLean (Macmillan, 1986). Nor was it my intention to offer an explanation of the arms race. Many factors are involved that are not discussed here. However, I hope that by examining how nuclear weapons decisions are taken in a comparative way, this book may contribute towards our understanding of the forces driving our nuclear-armed world. The nuclear weapons countries are not mirror images of one another. Perhaps by improving our knowledge of how they come to these decisions, we may better understand how to introduce restraint.

I would like to thank all my colleagues in the Oxford Research Group for their friendship and support. Scilla McLean suggested the theme of this book and co-ordinated the research on which it is based. She gave guidance and editorial support throughout. Tony Thomson provided invaluable help with research. I am grateful to John Hamwee and Margaret Blunden, who read the entire text and made useful suggestions, and to Janet Dando, for her painstaking subediting. John Beyer, Julian Cooper, Macdonald Graham, Francois Nectoux, Andrew White, Nancy Ramsey, and David Schorr contributed to the background studies on which this book is based. The first five commented on drafts and gave me helpful insights. I am also grateful, for discussions and comments, to Nigel Dudley, Michael Flood, Sean

Gregory, Pauline Hodson, Simon Hodgkinson, Bruce Kent, Claire Leggatt, Simon Lunn, Clive Ponting, John Poole, Justin Rosenberg, Josie Stein, Jim Steinberg and Marjorie Thompson. Officials in the Arms Control and Disarmament Research Unit of the Foreign and Commonwealth Office and in the Ministry of Defence provided helpful information. I would also like to thank the many unnamed journalists and researchers on whose stories and observations I have drawn. I alone, however, am responsible for errors of fact and interpretation.

Financial support to the Oxford Research Group during the period when this book was written was generously provided by the Joseph Rowntree Charitable Trust, the Barrow and Geraldine S. Cadbury Trust and the Resource Group.

London   HUGH MIALL

1 Introduction

[Cruise missiles – SS20s – Chevaline – Lop Nor – Mururoa – the decision-makers behind the deployments]

Early on the morning of 14 November 1983, on a cold, grey, autumn day, a heavily-laden US C141 Starlifter aircraft broke through the clouds above Berkshire in southern England and began its landing. Underneath, in the market town of Newbury, people were beginning their day's work and children were arriving at school. Outside the air-base, women protesters, clustered around their makeshift shelters, were surprised by the plane's arrival a day earlier than expected. At 9 a.m., the plane touched down behind a double wire fence and taxied towards an inner security fence surrounding six half-finished concrete silos. As the plane came to a stop, helicopters hovered at either end of the runway and British paratroopers and US Air Force men stood guard. Two large crates covered in tarpaulins were hauled out of the plane's cargo bay and rolled away on trailers.

Each crate contained a missile. Shaped like a torpedo, the missiles were 21 ft long and 21 in. in diameter. They weighed 2650 lb. At the back, they had stubby 4-ft wings, and a jet exhaust nozzle. At the front they had a round nose cone. Inside the cone was a 3-ft space for the warhead. Its code name was W-84.

If it was ever detonated, the warhead would explode with over ten times the destructive force of the bomb which devastated Hiroshima.

One of the women wept. 'I never thought it would happen', she said. 'I didn't believe they would fly in the face of world opinion.' Several others were close to tears.

The Secretary of State for Defence, Mr Michael Heseltine, cut short a visit to Aldershot and returned to London. There he met the Prime Minister, Mrs Margaret Thatcher. Later he went to the House of Commons to inform MPs. In accordance with a NATO decision, he said, the first flight of Ground-Launched Cruise Missiles had arrived in Britain. The first missiles would become operational by 31 December and a total of 160 would be deployed at the US air-bases at Greenham Common and Molesworth. 'For many people, Greenham Common has become a symbol', said Heseltine, at an earlier visit to the base. 'For me too it is a symbol, a symbol of NATO's determination to ensure the continuing success of its policy of deterrence.'

1

Later that day Mrs Thatcher gave an address at the Lord Mayor's Banquet in the Guildhall, London. 'Of course we would rather not have these weapons,' she said, 'but the Soviet Union has tried to insist on a monopoly of modern medium-range missiles in Europe. That monopoly we simply cannot accept. . . . We really mean it when we say that we want to negotiate arms control agreements, not only on nuclear weapons but on conventional forces as well . . .'. 'We will do everything possible to reduce the risks of war, and to avoid the misunderstandings which increase those risks. Britain is ready to pursue, in the right circumstances, a sensible dialogue with the Soviet Union and the countries of Eastern Europe. We want, and will work for, a safer world. Let it never be said that we failed because an East and West misunderstood one another.'

*   *   *

Two months later, the East replied to the West. In the northern Baltic state of Estonia, some 1200 miles to the east of Greenham Common, local residents watched as two batteries of mobile missiles were deployed. 'We now have definite information that two two batteries of SS5s have been swapped for SS20s', said Mr Julijis Kadelis, a spokesman for the Baltic World Conference. Speaking in Stockholm, he quoted eyewitness reports and Western satellite photographs, which located the mobile batteries near the towns of Tapa and Koppu. Local residents were concerned about the siting of the missiles in the area, he said. In 1968, a missile and an ammunition plant had exploded, killing an unknown number of people.

The SS20 is a two-stage ballistic missile. It is 52 ft long and weighs 25 tons. Its range is about 3000 miles, sufficient to reach all parts of Europe, but not the USA. It has three warheads, each ten times more powerful than the Hiroshima bomb.

SS20 missiles had first been deployed in the Soviet Union in 1977. By 1982, 243 had been deployed facing Europe, and 126 facing China. But the deployments were halted in 1982 by General Secretary Leonid Brezhnev, in a unilateral moratorium. The Baltic States had, up till then, been free of SS20s. Now the Soviet authorities were resuming deployments, and moving the missiles forward to the Baltic.

The Soviet government had warned in 1983 that it would treat the deployment of Cruise and Pershing missiles as an 'additional threat' which it would meet with more weapons.

In a statement, the British Foreign Office said it was 'preposterous' to suggest that the SS20 was justified as a counter-weight to NATO deployments, 'given the number of Soviet intermediate-range missiles already targeted on NATO territory'.

*   *   *

The northwest coast of Scotland is a rugged and beautiful landscape, with deep lochs, scattered fishing villages and small farms. The ancient hills of the Grampians sweep down to the sea in graceful lines. In the summer, travelling across the Firth of Clyde in a ferry, one can see, on a clear day, a vista of green mountains and sparkling blue water. In the autumn, however, mist often hangs over the Firth, casting a silent veil over the dark surface of the water. In such a mist, the unwary can be surprised by suddenly encountering a looming shape with a tower in the centre of it. This can be mistaken for land, but it moves, and if it comes closer, the swirl of water around its bow and the characteristic shape of a conning tower indicate the sinister profile of a submarine.

In November 1982, from such a ferry, it might have been possible to detect a submarine putting to sea from Gare Loch. The submarine was *HMS Renown* and it was making the first patrol with a new type of ballistic missile called Chevaline. The following small item appeared in the International Defence Review:

Chevaline enters RN service. The British Ministry of Defence has formally announced that the Chevaline submarine-launched ballistic missile has entered service. Chevaline, officially designated Polaris A3TK, will gradually replace the 16 Polaris A3s carried by each of the Royal Navy's four SSBNs. The Chevaline "front end" is programmed to manoeuvre in space, releasing both nuclear warheads and large numbers of dummies.

*   *   *

Since the first mushroom cloud arose over Lop Nor in the remote and uninhabited Gobi desert, this area of Xinjiang province began to attract the attention of the world. In the mid-summer of 1984, Guo Cheng visited China's nuclear testing ground.

I departed by car from a place called Malan and headed for the ruins of the ancient city of Loulan. I happened to travel in the same car with Zhang Zhisan, an old friend of mine whom I had not met for a long time. The former commander of the nuclear testing base, Zhang Zhishan told me that the Lop Nor nuclear testing ground, with a total area of more than 100,000 square kilometres, is as large as Zhejiang Province. Nowadays in the testing zone there is a highway network with a total length of more than 2000 kilometres as well as various facilities for nuclear tests to be carried out on the ground, on towers, in the air, by missiles, in horizontal underground tunnels or in vertical shafts . . .

We drove for many hours along the road which oozed "oil" under scorching sunshine. An odd scene finally presented itself before our eyes – dilapidated automobiles lay on rocks, armoured cars and aircraft had been turned into wreckage, dilapidated concrete buildings could be seen here and there and part of the surface looked like melted glaze. It looked as if a large-scale modern war had been fought in the depths of this desert some time ago! . . .

During my visit to the test areas people there were busy preparing for another nuclear test. Trucks carrying experimental equipment were speeding along the high roads; scientific and technical personnel were sweating away at their work in tents; and leading cadres of the base and various sections were giving commands at the test site. The quiet site was bustling with activity.

At the mouth of the shaft people were rehearsing the installation of an "experimental product". I saw that several base leaders in charge of technical work were directing the work there. All technicians were working with rapt attention. It seemed that they were handling a real nuclear bomb rather than a dummy one.

On 3 October, 1984, a military observatory at Hagfors in western Sweden recorded an underground nuclear test in China, measuring 5.7 on the Richter scale.

Two days before this test, on 1 October, the Chinese National Day Parade took place in Beijing. Schoolgirls decorated in garlands of flowers marched through the streets, alongside an enormous bust of Mao Zedong. Also in the parade was an array of military hardware, mostly based on the technology of the 1960s. One intercontinental missile, about 100 ft long, rumbled through the city.

In a speech given above the Tien An Min Square, Deng Xiaoping called for the armed forces to strengthen the national defences 'in the seriously deteriorating international situation'. Deng went on to say that the unification of China, including Taiwan, 'is rooted in the hearts of all descendants of the Yellow Emperor'. 'We stand for settlement of international disputes through negotiation', he said.

Twenty years earlier, Deng Xiaoping had been responsible for sending Zhang Aiping (later the Minister for Defence) to direct the Chinese bomb programme. 'Have courage and go ahead', Deng had said to Zhang. 'If you succeed, the credit is yours, and if you fail the secretariat will take the blame.'

China's first missile, the 'East Wind', was tested in 1966. The largest and most recently deployed Chinese missile, the CSS4, was first tested in 1980. Its range is 7000 miles and it is capable of reaching Moscow and the Western United States. It is believed to carry a single 3- to 4-megaton warhead. Only two to four of the missiles are in service.

China now has about 100 nuclear missiles, mostly of medium range, as well as about a hundred nuclear bombs for air delivery. Submarine-launched ballistic missiles and tactical nuclear weapons are under development.

*   *   *

Mururoa Atoll lies between Pitcairn Island and Tuamotu in the South Pacific. It is a long, narrow strip of land, dozens of miles long but only a few hundred yards wide at its widest point. Remote and isolated, its blue-tinged lagoon is lined with white beaches and waving coconut palms.

Mururoa was uninhabited before 1964. Apart from an occasional ship, no people disturbed its tranquillity.

Now the atoll has a population of over 3000. Eighty per cent of them are male, and all are government officials. They live in a pre-fabricated barracks at the centre of the atoll. They have a television station, a radio station, two cinemas, and a hospital. All their food has to be imported by ship from Tahiti. A bakery on the atoll bakes a tonne of bread every day. There is a desalination plant to provide water, though the inhabitants prefer to drink imported mineral water, and consume 6000 litres of it a day.

They work in modern buildings and laboratories which are full of sophisticated equipment for measuring radiation, blast and seismic waves. They build roads along the atoll and operate drilling equipment which is run 24 hours a day to bore holes into the basalt rock beneath the surface. They are all employees of a single organization, the Pacific Experimentation Centre. They all work to a single purpose, dictated by officials of the Directorate for Nuclear Testing Centres, an arm of the Commission d'Energie Atomique, based in Paris, 11500 miles away.

Over forty nuclear bombs have been exploded in the atmosphere at Mururoa, the last in 1974. A further forty have been exploded underwater and underground. The subsidence caused by these detonations has made the ground level of the atoll drop by a metre and a half. Cracks have opened in the coral and rock beneath the atoll, and radioactive material from the tests is leaking through these cracks into the sea. The subsidence causes small tidal waves. To counter them, the Pacific Experimentation Centre has built a series of 6-ft high concrete walls along the atoll and installed a monitoring system, warning alarms and 20-ft high safety platforms.

In July 1982, five successive underground explosions were carried out at Mururoa. France was testing the neutron bomb.

*   *   *

These are only some of the more conspicuous instances of a continuous process of deployment and testing of nuclear weapons carried out every year by the United States of America, the Soviet Union, Britain, France and China. On average, one nuclear test is conducted every week. Two new missiles are deployed every day. Only a small proportion of these deployments and tests attracts the attention of the media.

The deployment of weapons and testing of warheads are themselves only the most visible stage of a much larger process. Before a new weapon is deployed, it has to be designed, planned, financed, approved, and manufactured. Long before the decision is taken to deploy, critical decisions are taken at the research, development, procurement and production stages.

A nuclear weapon consists of three parts – a platform, a delivery vehicle and a warhead. The platform may be a submarine, ship, aircraft, vehicle or silo. The delivery vehicle may be a missile, bomb, artillery shell or depth charge. The warhead may be in a single or multiple configuration, ranging in explosive yield from as little as 0.01 kilotons to over 20 megatons.

The production of a complete, modern, nuclear weapon system is an exacting task, even for an advanced industrial state. Typically, it takes ten to fifteen years from concept, through assembly line, to deployment. The warhead, delivery vehicle and platform will usually be developed separately and brought together to form the new weapon.

To fuel the warheads, enriched weapons-grade uranium or plutonium is required. This is produced in uranium enrichment plants and nuclear reactors. The nuclear materials are then assembled into warheads at purpose-built laboratories and plants. The warheads for the cruise missile were produced at the Y-12 plant at Oak Ridge, in the foothills of the Great Smokey Mountains in Eastern Tennessee. Soviet warheads are fabricated at Chelyabinsk in the Urals. The British warheads are made at the Atomic Weapons Research Establishment at Aldermaston, in the undulating wooded countryside of Berkshire, and at its associated plants in Burghfield and Cardiff.

The scientists who work in the nuclear weapons laboratories design the warheads and provide advice to the defence ministries and military services. In the USA, top nuclear scientists sit on the Science Advisory Boards which the Department of Defense consults. In the Soviet Union, the scientists sit on Scientific–Technical Committees of the Defence Ministries and Armed Services, which screen and evaluate proposals for new weapons.

To produce the missiles, large, technically-sophisticated plants are required, such as the laboratories and factories of the aerospace corporations. Frequently their engineers are working on tasks which lie at the limits of existing knowledge. The cruise missiles were made at the Convair Division of the General Dynamics Corporation in San Diego, California. The SS20s were designed at the Nadiradze Design Bureau near Moscow. These plants are as integral to the nuclear strength of a state as the missiles in their silos.

In the United States, the defence contractors who manufacture the missiles, aircraft and submarines which carry nuclear weapons are private corporations. Dependent on military contracts, they compete for business by seeking to out-perform one another in military technology. Corporations like Boeing, General Dynamics, Grumman, Lockheed, Rockwell, McDonnell Douglas and Northrop lobby the Congress and the Defense Department, and exchange personnel with the military and the government. In France, the main defence contractors are nationalised, and closely linked with the powerful Delegation Generale pour l'Armament, the procurement arm of the Ministry of Defence. In the Soviet Union and China, where these corporations have no direct equivalents, the functions they perform are carried out by bodies within the government, such as the Design Bureaux of the Defence Industry Ministries. Leading designers are consulted by members of the Politburo over nuclear weapons decisions and can be personally influential.

The armed services of the five powers are the customers for the nuclear weapons. Both the United States and the Soviet Union have specialised strategic forces: the Strategic Air Command of the US Air Force, and the Soviet Strategic Rocket Force, which was formed from artillery units. These services command the land-based intercontinental missiles. The Armed Services of the United States and the Soviet Union have their own weapons procurement arms and exert a considerable influence over the selection of new weapons. Britain, France and China also have nuclear commands, and those who have risen through them are now to be found in decision-making posts of key importance for the acquisition of new weapons.

The 10 to 15-year gestation period of a nuclear weapons system is longer than the lifespan of a government or administration in the western democracies. Consequently, much is left to the officials in the defence ministries. A battery of high-level committees, most of them little known and operating in secret, prepare and take important decisions in all countries. Permanent, unelected officials may be in posts long enough to manage most or all of the life of a weapons project. In France and Britain, they sometimes acquire more real power and influence than the Ministers they serve, who frequently hold office for short terms. In the United States, where new Administrations appoint officials to take the top jobs in government departments, and in the Soviet Union where the Party closely supervises the government, the situation is different.

At the highest levels of government are the Cabinets, Politburos, Prime Ministers, Presidents and Party General Secretaries. Immediately beneath them are high-level bodies which operate the central machinery of government, such as the Cabinet Secretariat in Britain and the Central Committee Secretariat in the Soviet Union. There are also top-level councils where political leaders consider defence matters, such as the French Conseil de Defense, the Soviet Defence Council and the Central Military Commission of China. In the Soviet Union and China, many leading figures with responsibilities for the development of nuclear weapons are also members of either the Politburo or the Central Committee. In the United States, in contrast, the President has no regular role in planning new nuclear weapons.

In order to reach the stage at which it is a candidate for deployment, a new nuclear weapon in the West must have won the approval of some or all of these groups. It must be technically 'sweet', it must have the backing of a defence contractor, it must be accepted as militarily useful by one of the Services, it must win the approval of the defence officials, and the high-level agencies must endorse it. To have reached this stage, a weapon will have had a considerable amount of money invested in it, and will have built up an enormous momentum in the support of vested-interest groups.

A decision by a Government to kill the development of a weapon at the deployment stage takes exceptional determination and investment of much political capital. Yet this is sometimes the only stage in the weapon's development in which the democratic mechanisms are fully engaged. In the UK, for example, Chevaline was developed for 13 years, at a cost of £1000 million, before Parliament was informed.

None of the five major nuclear weapons states consulted their respective parliamentary or other representative institutions before acquiring nuclear weapons. The US Congress, the British Parliament and the French Parlement all vote through the funds which are used to build nuclear weapons. How much control do they have over the development of new weapons, and how effectively do they exercise this control? Although their constitutional position is different, the National People's Congress of China and the Supreme Soviet of the Soviet Union are indirectly-elected bodies. What role do they play?

The most effective form of restraint on the development of new weapons is often budgetary stringency. But nuclear weapons are given high and sometimes overriding priority. How are the nuclear weapons programmes financially controlled and how effective is this control?

It is the people of the major nuclear weapons states in whose names the nuclear weapons have been built. What has their opinion been? What effect has public opinion had on decisions to develop nuclear weapons?

If the arms race is to be brought under control, the nuclear states will have to agree, either independently or in concert, to halt and then reduce the development of new weapons. The main forum for international discussions to seek agreements to limit or reduce nuclear weapons has been the bilateral negotiations between the United States and the Soviet Union, though all five nuclear states are represented at one multilateral forum, the Conference on Disarmament at Geneva. All five nuclear states profess a desire to reach agreement on multilateral limitations to arms, yet all five are at the same time engaged in efforts to modernise their arsenals and make them more militarily effective. What is the relationship between decision-making for new nuclear weapons, and decision-making for arms control? How are arms control and disarmament positions formulated?

Nuclear weapons represent the most awesome concentration of power at the disposal of any human group. If they are to be brought under restraint, it is necessary first to examine how the decisions to develop them are taken.

2 The Weapons Laboratories

[Nuclear warheads – Aldermaston – Livermore – The Ministry of Medium Machine-Building – Wang Ganchang and the Chinese nuclear programme – the CEA – the influence of the nuclear scientists on government policy and on the test ban talks]

Inside the warhead of every nuclear weapon, there lies a machined sphere of plutonium inside a larger sphere of beryllium and uranium. This is enclosed in an outer shell of conventional explosive, punctured by radially mounted detonators. Near these concentric spheres lies a truncated cone of lithium deuteride, the hydrogen part of the bomb. On detonation, the beryllium and uranium would be driven by the conventional explosive into the plutonium core, starting a chain reaction in the plutonium. The energy generated in the fission of the plutonium would then be focussed on to the lithium deuteride, raising its temperature to that of the sun and triggering a more powerful fusion reaction in the lithium deuteride. During the arming sequence, tritium and deuterium are pumped into the bomb, the precise amount controlling the device's destructive yield.

The scientific and technical skills and resources required to maintain these nuclear warheads and to design new ones is concentrated in a handful of nuclear weapons laboratories around the world. These are, in Britain, the Atomic Weapons Research Establishment at Aldermaston; in the United States, the Los Alamos National Laboratory, the Lawrence Livermore National Laboratory and the Sandia Laboratory; in the Soviet Union, the Institutes and design laboratories of the Ministry of Medium Machine-Building, thought to be in the Chelyabinsk region in the Urals, and elsewhere; in China, the Institute of Atomic Physics in Beijing and the design facilities of the Ministry of Nuclear Energy at Lanzhou in Gansu; and in France, the 'centres d'Etudes' of the Direction des Applications Militaires of the Commission d'Energie Atomique at Ripault and elsewhere.

These laboratories contain some of the most closely-guarded secrets of the nuclear powers. Their leading scientists and administrators, who are privy to information which reaches only a narrow circle with the highest security clearance, play a prominent part in decision-making on nuclear weapons.

Basic and applied scientific research is a key stage of nuclear weapons development, since it is on scientific and technological innovations that new weapons systems depend. The politicians, defence ministry officials and military can suggest guidelines or targets for this research, and by funding particular lines of research and not others can influence its direction. But it is the scientists themselves, and the underlying momentum of technical innovation and scientific discovery, that determine which new weapons are conceived.

*   *   *

The Atomic Weapons Research Establishment at Aldermaston is both a nuclear weapons laboratory and a production facility. It designs, develops and tests warheads, not only for the British submarine-launched missiles, but also for the Royal Air Force's nuclear-capable aircraft and for the Royal Navy's nuclear depth-charges. Since the warheads contain radioactive elements which decay over time, they need regular maintenance and this is carried out at Aldermaston. The nuclear scientists are responsible for the design of new warheads for new weapons and for keeping abreast of the 'state of the art' by pursuing their own original research in warhead design. They also manage the special collaborative arrangements for nuclear co-operation and transfers of nuclear material with the United States.

With some 5000 employees and an operating budget of around £100 million a year, the Aldermaston complex is a sizable establishment. It is located in a former airfield on the flood-plain of the Thames Valley. Once a virtually independent scientific fiefdom, Aldermaston's activities are now co-ordinated by the Ministry of Defence, with the Director of Aldermaston holding a dual post as Deputy Chief of Defence Procurement (Nuclear). Aldermaston's director is also responsible for the control of the two Royal Ordnance Factories involved in the production of nuclear weapons – ROF Llanishen at Cardiff and ROF Burghfield near Reading.

Aldermaston is concerned with the 'physics of extremes'. In order to study conditions at the centre of a nuclear explosion, scientists examine what happens to matter at temperatures of 100 million degrees centigrade and pressures equivalent to 20 million atmospheres. These are simulated in the laboratory through the use of one of the most powerful lasers in Britain, the most powerful light-water research reactor and the only fast-pulse reactor in Europe. This reactor, named 'Viper', produces intense pulses of neutrons and gamma rays with a peak power of 20 000 megawatts. These facilities have been built to test the ability of nuclear warheads to survive a nuclear explosion set off by an anti-ballistic missile defence.

Other facilities include the most powerful computer in the country, a hot-metal rolling plant, a plutonium-handling plant, a uranium plant and an Apprentice Training School for 160 school-leavers.

The people who work in this environment wear white coats or jackets and ties. They are diligent, precise and thorough. Their faces express concentration and patience. They watch computer monitors, peer down microscopes, and make delicate adjustments to warheads with specialised tools.

The director of the Atomic Weapons Research Establishment since 1982 has been Peter Jones. Born in 1925, Peter Jones' early studies in engineering were interrupted by the war, when he flew for the Royal Air Force. He then took a physics degree at London University and worked for the General Electric Company for several years before starting work on the British nuclear warhead programme in 1954 at Fort Halstead. He later moved to Aldermaston to work in the weapons system trials division and led the weapon measurements group throughout the series of thermonuclear tests at Christmas Island. After returning to Aldermaston, he became the head of the warhead electronic firing systems division in 1966. He conceived the Chevaline project in 1969 and co-ordinated its development at Aldermaston from 1969 to 1980. In 1974 he was promoted to Chief of Warhead Development, in 1979 he became Principal Deputy Director and in 1982 Director of Aldermaston. He lives near Basingstoke with his wife Jackie and two sons and a daughter. He has an active interest in flying and is a keen motorist, specialising in Cadillacs. Jones has gold-rimmed spectacles and piercing blue eyes beneath a high forehead and swept-back hair. He relishes his former title of chief warhead designer – 'there were only six of us in the world'.

Jones sees the prime motive of the weapons scientists as technical achievement. 'The motivation for most technical people is that they can achieve something . . . the real motivation behind technical people is still achievement.' He stresses the importance of retaining an 'industrial base' at Aldermaston. 'First of all there is the government requirement that the country shall remain a nuclear power. That requires a base capability and we have to maintain that. Superimposed on that is the requirement that the government may have from time to time for a particular project. It is that, like Trident . . . which maintains the project disciplines and keeps those capabilities up to scratch.'

The purpose of building Chevaline was to ensure that British nuclear warheads would be capable of destroying Moscow. In 1962 American intelligence observed the beginnings of construction of an anti-ballistic missile system around Moscow. Khrushchev boasted that the Soviet Union had an ABM that could 'hit a fly in outer space'. By 1968 US intelligence believed that the system could supply a limited defence of the Moscow area. Peter Jones recalls: 'By that time we realised that it was very significant in terms of the effect on Polaris and that we would have to take any changes we made to Polaris very seriously indeed.' After consultations with the inner Cabinet, the Prime Minister, Harold Wilson, asked the Atomic Weapons Research Establishment to undertake serious studies into ways of improving the Polaris missiles.

Studies began under Peter Jones' direction, and in 1970 Aldermaston was authorised to start a feasibility study on a system which would incorporate multiple warheads, decoys, and hardened warheads to withstand radiation. Work on the project continued at Aldermaston, under the authority of the Chief Scientist at the Ministry of Defence, until 1976. During this period the nuclear scientists organized the contracts to industry for studies and components, and took all the key design decisions.

Considerable uncertainty surrounded the future of the project. At times the Navy, and the Heath Government, appeared to prefer to buy Poseidon missiles from the USA as an alternative to developing Chevaline. At times there was doubt over the extent of the ABM threat, especially after the 1972 ABM Treaty which limited the USA and the Soviet Union to one system each. The project was 'trickle-funded' for periods of three and six months. Nevertheless, on every occasion that the project was brought before the inner Cabinet – which happened only rarely – the politicians accepted the advice of the nuclear scientists to continue.

A crisis occurred in 1976 when it became clear that the costs of the project were rising well above original estimates and that the difficulty of designing a multiple re-entry vehicle had been underestimated. Moreover, development of the ABM system around Moscow had stopped and military opinion now doubted that the system could ever have posed a serious threat even to the original Polaris rockets. In spite of this, the inner Cabinet again decided to proceed – although the control of the project was taken out of Aldermaston's hands. The primary reason to continue was to maintain the work programme at Aldermaston, in order to avoid the dispersal of the design teams and consequent inability to build new warheads in the future.

The directors of Aldermaston have persisted in presenting the government with a choice between developing new nuclear weapons and the gradual atrophy of Aldermaston's capabilities to make new weapons. No British government has been prepared to countenance the second option, so the pressure for new weapons for Aldermaston has been a powerful imperative. Recently Colin Fielding, Peter Jones' predecessor as Director of Aldermaston and now Controller of Research Establishments (Nuclear) in the Ministry of Defence, said in evidence to the House of Commons Select Committee on Defence that, if the Trident programme was cancelled, there would be a need for 'close maintenance of our capability if there was ever any intention to resume design and development of nuclear warheads ever again'.

*   *   *

The warhead for the Ground-Launched Cruise Missile was designed at the Lawrence Livermore National Laboratory in Livermore, 40 miles east of San Francisco. This laboratory was set up in 1952, to spur the development of the hydrogen bomb, following the violent disagreement between Robert Oppenheimer and Edward Teller over whether it should be developed. Teller won the argument and his new laboratory made the H-bomb.

All the nuclear warheads in the US arsenal are designed either at Livermore or at Los Alamos. Los Alamos lies in the Jemez mountains northwest of Santa Fé in New Mexico. Its Spanish name means 'the Poplars', after the trees which give the area its great natural beauty. A third laboratory, Sandia, in Livermore, is responsible especially for the engineering of nuclear weapons design, non-nuclear components, and control of maintenance of the stockpile of nuclear warheads.

The great majority of new warheads are modifications of existing designs, with the explosive yield, size, or shape altered to fit into a new weapon. The W-84 for the Ground-Launched Cruise Missile was a modification of the B-61 light-weight bomb. The work was assigned to Livermore in September 1978. Work on the warhead for the Air-

Launched and Sea-Launched Cruise Missiles was assigned to Los Alamos.

The government stipulates three phases in the development of a new warhead design. Phase 1 is the definition of a weapons concept, which can be initiated by any organisation or contractor to the US Department of Energy or the Department of Defense. Phase 2 is a feasibility study. This has to be approved by the Director of Defense Research and Engineering, a key decision-maker in the Department of Defense, on the basis of a joint study carried out by the Department of Defense and the Department of Energy, co-ordinated by a Military Liaison Committee within the Department of Energy. Phase 3 is authorisation of the development of the warhead. This has to be made by the Secretary of Defense. Within 60 days of Phase 3 approval, a letter giving details of this together with other changes in the stockpile is sent to the President, who signs it. He thereby authorises the development, production and retirement of every US nuclear warhead design.

The Cruise W-84 warhead received its Phase 3 clearance routinely in January 1979, three months after the project was assigned to the laboratory. Development work then continued until late 1983 – the time of the deployment of the missiles in Europe – when the warhead was deployed. It became operational in December 1983.

By the time the W-84 was designed, nuclear warheads of this kind had become a mature technology. The plutonium at the centre of 'Fat Man', the bomb which was dropped on Nagasaki killing over 40 000 people in the first few seconds, was about the size of a grapefruit. But the bomb itself weighed five tons. The hydrogen bomb which Teller and his colleague Ulam designed was more than a thousand times more powerful than the atom bomb. Consequently it was possible to make nuclear weapons both smaller and much more destructive.

In the late 1950s and 1960s further refinements were made to increase the ratio of destructive yield to weight, so that several warheads could be placed in the tip of one missile. The Cruise warhead is the culmination of a long programme of miniaturisation.

Livermore scientists also designed the warheads for the Minuteman II, Minuteman III, Poseidon and MX missiles.

Most of this work was incremental – building 'follow-on' weapons from proven designs. However, an important part of the task of the weapons laboratories is to explore qualitatively new weapons concepts. 'It is the laboratories' job to propose new alternatives for discharging military missions', writes Michael May, the current associate director of the Livermore Laboratory. 'This is sometimes welcomed by the military services, sometimes opposed by them. It may be said of this activity that the laboratories are driving the 'qualitative arms race'. . . . We cannot avoid technical change', he goes on, 'nor can we expect that the laboratories will not play a role in this change.'

One example of a qualitatively new concept was the neutron bomb – a nuclear weapon designed to release more of its energy as radiation. This was conceived at Livermore in 1959. No political or military requirement had been set; the neutron bomb was simply another technical possibility for a nuclear weapons application which the laboratories continuously explore. For many years the directors of the laboratory promoted the neutron bomb without success. 'I knew we were pushing hard', recalls a Livermore physicist who worked on the neutron bomb. 'That was one of our things to sell for many years.' Teller and his colleagues persistently argued against the Limited Test Ban on the grounds that it could foreclose the development of neutron warheads as precise battlefield weapons. Although the weapons scientists were blocked for many years by the lack of military enthusiasm for these weapons, in the end Livermore did develop neutron bomb designs for production for deployment in artillery and missiles in Europe.

Livermore is now heavily involved in the design of a new 'third generation' of nuclear weapons. The first generation was the atomic or fission weapon. The second was the hydrogen or fusion bomb. The third generation is described as 'weapons in which a part of the total energy produced by the explosion is converted in some way to a form more precisely tailored to the need than just blast and heat'. The best-known example of a third-generation weapon is the X-ray laser, which converts part of the energy of a nuclear explosion into a precisely-directed beam of X-rays. This may have potential applications as a space weapon.

The basic theoretical work on these new weapons is conducted at Livermore by young physicists in their 20s and 30s. They work in jeans, checked shirts and running shoes. Many have beards. They drink Coke and visit ice-cream parlours to relax. Many of them were recruited through the Hertz Foundation, a trust started by John D. Hertz, whose name is better known in association with the car-rental business.

One of these Hertz scholars is Peter Hagelstein, a brilliant physicist trained at MIT. He ran marathons at college and played violin in a string quartet. He wanted to develop the first X-ray laser, for which he saw applications in medicine and biology. In the 1970s he worked 14 or

15 hours a day to develop computer codes to predict the electron transitions of laboratory X-ray lasers. He joined Livermore to pursue this work, not intending to do weapons work. (About half of the Livermore laboratory's work is in non-weapons areas.) Indeed, Hagelstein consciously avoided weapons applications. But another scientist, George Chapline, was working on X-ray lasers and had proposed a novel way to build one with a nuclear bomb as its power source in 1977. During a meeting with Chapline, Hagelstein inadvertently let slip a suggestion which radically improved Chapline's device. Hagelstein was then asked to use his computer codes to work on the X-ray laser weapon. 'I got my arm twisted to do a detailed calculation. I resisted doing it. There were political pressures like you wouldn't believe.' Teller persuaded him forcefully: 'The Force has a powerful effect on the weak mind', said Hagelstein, quoting self-effacingly from the film 'Star Wars'. During this period, too, Hagelstein read Solzhenitsyn's *The Gulag Archipelago*. Despite his own misgivings, Hagelstein eventually agreed. Later he became progressively more involved with the weapons work. 'My view of weapons has changed', he reflected. 'Until 1980 or so I didn't want anything to do with nuclear anything. Back in those days I thought there was something fundamentally evil about weapons. Now I see it as an interesting physics problem.'

With the theory of the X-ray laser proven in tests at Nevada, Teller began to use his formidable persuasive powers on the federal government. First an article about the possibilities of space weapons was leaked to the influential *Aviation Week*. Teller also worked through the Heritage Foundation, a right-wing think-tank which included people close to Reagan. He wrote an article in *Reader's Digest* (the President's favourite magazine) and spoke to George Keyworth, Reagan's science adviser and his own former protégé. There followed Reagan's 'Star Wars' speech of 23 March 1983, in which, to the astonishment of the US Department of Defense, the President called on the scientific community 'who gave us nuclear weapons to turn their great talents to the cause of mankind and world peace – to give us the means of rendering those nuclear weapons impotent and obsolete'. In the controversy which followed, much criticism was made of the feasibility of this intention, but all sides of the political establishment unified around the apparently uncontentious proposition that vigorous research should go ahead. Teller's lobbying had been spectacularly successful.

All the basic skills, technology and resources for designing nuclear weapons in the USA are concentrated in the weapons laboratories. The government has avoided sharing the technology with other institutions. Consequently, those in the weapons laboratories have unique expertise, and unique access to secret information. By the same token, those who enter this field cannot readily take their skills to other environments, without abandoning weapons work. As a result, there has built up within the laboratories a large pool of scientists and technicians who have been involved in making key inventions, undertaking critical research and advising governments over many years. This group has acquired great influence. The directors and leading staff of the laboratories have served on most of the Science Advisory Boards, which advise the Services on new weapons and on many Presidential advisory committees. Some of those who have left the laboratories have taken senior decision-making posts in the government. George Keyworth, then President Reagan's science adviser, was a nuclear physicist at Los Alamos. Dr Gerold Yonas, who used to be a group head at Sandia, is now chief scientist to the Strategic Defense Initiative. Herbert York, Livermore's first director, became Director of Defense Research and Engineering at the Pentagon (the key post in selecting new nuclear weapons systems). Harold Brown, the second director, also became Director of Defense Research and Engineering, then Secretary of the Air Force, and then President Carter's Secretary of Defense. Duane Sewell, a deputy director, became Assistant Secretary of Energy. Richard Wagner, Associate Director at Livermore until 1981, is the present Assistant Secretary of Defense for Atomic Weapons.

Such a group, isolated and to a degree ostracised by scientists in civil fields, with a wall of secrecy between its activities and the outside, almost inevitably develops its own set of values and assumptions. These can be traced to some extent in the public statements of the American scientists.

A basic view held by most of the weapons scientists is the need for deterrence. Michael May, associate director of Livermore, writes: 'What is this qualitative arms race? If it is a process of replacing older systems by new ones which will do a better job of strategic deterrence, or of tactical deterrence in Europe, or of strategic defense – assuming this to be a desired addition to our posture – then we accepted that process when we decided to accept armed deterrence in the first place.' He goes on: 'To most people in the US and Soviet defense establishments, and probably to most US and Soviet citizens, the weapons symbolize an ultimate recourse, one that everyone hopes will never be used, but which the governments see, rightly or wrongly, as necessary to discharge the very responsibilities which their citizens lay upon them.'

'Nuclear weapons were bound to come', says May, in a chilling passage. 'Our physics world and our political world have conspired to bring them about, if not at Alamogordo and Hiroshima, then elsewhere. Robert Oppenheimer and company were not any more responsible than anyone else for bringing in the atomic age. We all build atomic bombs, as we all build the institutions that demand them.'

Cory Coll is a fissions explosives expert at Livermore. 'There's no question that most people feel that if nuclear weapons had never been invented and never could be invented, we'd be better off', he says. 'Unfortunately, nuclear weapons have been invented, and they can't be uninvented; and we have to live with this threat and have to try to ensure that they are never used. We all have families, we all want to see grandchildren, we all want to think our grandchildren have attractive futures. We reach for easy solutions that we think are going to provide answers, and in this area, I think there are no easy answers; and there are certainly no answers that are going to happen quickly. This is something that in my own view we're going to have to continue to deal with, certainly through the rest of my life. I don't expect that by the time I die we'll have eliminated nuclear weapons from the face of the earth.'

George Dacey, President of Sandia National Laboratories, was asked for his views on a nuclear 'freeze'. 'I think my answer is a very simple one', he said. 'If you mean by "freeze" that you intend to stop thinking, to stop considering what the weapons possibilities are, what modern warfare can, in fact, become, then I think you are taking a dangerous risk with this country's security. From a technical standpoint I think there are enormous possibilities for improvement ahead. We do need to make it clear that such progress is in the public interest and that we should charge on at full speed.'

*   *   *

The Chelyabinsk region in the Urals is thought to be the principal centre for the fabrication of Soviet warheads and the associated research and design. The responsible Ministry is the Ministry of Medium Machine-building (Minsredmash). Of all the Soviet industrial ministries, this is the one about which least is known. It was established in 1953. Prior to this the nuclear weapons programme had been managed by a body called the First Main Administration of the Council of Ministers, under the overall control of L. P. Beria, the head of state security. Since 1957, the minister of Medium Machine-building has been E. P. Slavskii, who has enjoyed a remarkable continuity of leadership. Born in 1898, he was associated with the nuclear programme almost from its inception. The Ministry is responsible for the production of warheads and for the processing of nuclear materials for military purposes throughout the nuclear fuel cycle. It also runs the Soviet uranium mines.

Theoretical work on the first fission bomb was under the direction of Igor Kurchatov. A secret 'Laboratory No. 2 of the Academy of Sciences' was set up outside Moscow, in an area of potato fields. By the end of the 1940s the laboratory had become a small town. In the 1950s it was renamed the Institute for Atomic Energy. This Institute, which now falls under the State Committee for the Utilisation of Atomic Energy, is still the principal Soviet nuclear research centre. Its director since the 1960s, A. P. Aleksandrov, is the President of the USSR Academy of Sciences. It is one of a number of Institutes which pursue high technology research. These Institutes do not work exclusively on weapons but undoubtedly, some of their work is of military relevance, for example, work in the areas of nuclear physics, lasers, electronics, new materials, computers and control systems. The Moscow All-Union Research Institute for Inorganic Materials carries out research into plutonium and other nuclear materials, which is likely to contribute to the weapons programme.

There are two major centres for underground testing. One is at Semipalatinsk in the deserts of eastern Kazakhstan; the other is on the island of Novaya Zemlya, in the Arctic Circle.

From the earliest days, the Party exercised tight control. Beria was in political control of Kurchatov's group. B. L. Vannikov, the Commissar of Armaments and A. P. Zavenyagin, Beria's deputy, administered the programme. Under their direction a gaseous-diffusion plant was built in Siberia and plutonium production plants near Sverdlovsk in the Urals. Slave prison labour was used in the construction of the nuclear plants and in the uranium mines.

The Party leaders were rigorous in maintaining the pace. When the gaseous-diffusion plant fell behind schedule, Beria dismissed its director. Slavskii, the Minister of Medium Machine-building, had a particularly ferocious reputation as an industrial manager. Even an official Soviet publication described him as 'temperamental'.

Nevertheless, the scientists sometimes stood up to the Party. The leading physicist and Academician, Peter Kapitsa, refused point-blank to work on nuclear weapons, earning considerable displeasure from Stalin and later Khrushchev. Andrei Sakharov, who at the age of 32 was one of the designers of the Soviet hydrogen bomb, clashed sharply with Khrushchev over whether it should be tested. Sakharov petitioned Khrushchev, as Chairman of the Council of Ministers, to stop the test. 'As a scientist and designer of the hydrogen bomb, I know what harm these explosions can bring down on the head of mankind.' He also stressed the contamination of the atmosphere. Khrushchev declares in his memoirs that he was impressed by Sakharov's moral and humanistic considerations. 'However, he went too far in thinking that he had the right to decide whether the bomb he had developed could ever be used in the future.' Khrushchev referred to the suffering of the Soviet people in World War II, and declared that 'we can't risk the lives of our people again by giving our adversary a free hand to develop new means of destruction'. The test went ahead: the explosion was 57 megatons.

In December 1957 or January 1958, a catastrophic nuclear accident occurred at Kyshtm, in the Chelyabinsk region. Hundreds of square miles were contaminated. After symptoms of acute radiation sickness were found, the towns in the area were evacuated; huge quantities of food were destroyed. For a time this incident damaged the reputation of the atomic programme with the Party. Nevertheless, the weapons programme continued to be pushed forward.

Few details are known of the present attitudes of Soviet nuclear scientists, nor of their role in Soviet decision-making. The Academy of Sciences is still an important centre where information is exchanged and the Party can be influenced. Brezhnev used to talk to leading weapons scientists and rocket engineers at a Rest Home belonging to the Academy. Scientists sit on the Technical-Scientific Committees of the Ministry, which evaluate new projects. However, the key decisions are taken centrally by the Defence Council and the Politburo.

A key difference from the West lies in the way research in the Soviet Union is financed. Military-related R & D establishments are funded from the State budget, not on contract to the Defence Ministry or other defence departments. The development of new lines of research is thus controlled centrally, unlike in the USA, where innovations can be pursued independently by the laboratories on contracts which they themselves lobby to acquire. Since funds for advanced research do not depend on specific weapons development programmes, the institutes do not have a direct financial incentive to press for innovation in new designs. Soviet scientists are clearly as capable of innovation as those of the USA, but the Soviet institutional arrangements do not encourage the exploration and development of new weapon types in the restless, competitive, decentralised manner of the US laboratories.

*   *   *

Similarly in China, research and development on nuclear weapons is tightly controlled by the Party. The Party was quick to see the importance of nuclear energy, and dispatched a scientific delegation to Moscow to plan atomic collaboration only four years after the Communist revolution of 1949. A joint Sino–Soviet nuclear research laboratory was established at Dubna, on the Volga. A leading member of the Chinese team at Dubna was Wang Ganchang, now Director of the Atomic Energy Institute in Beijing. Wang had trained in the west and worked with Lise Meitner at the University of Berlin, shortly before she fled from Hitler's anti-semitic persecution. It was in this laboratory that Hahn and Strassman discovered, in 1938, that bombardment of uranium with neutrons produced a different element, barium. Lise Meitner had correctly interpreted the phenomenon as nuclear fission. At that time Wang was around 30. He later worked as a physicist at the University of California (under whose auspices the Livermore laboratory is now formally run). Then, from 1957 to 1959, he worked at Dubna with the Russians. In the 1960s he became scientific director of the Chinese bomb programme. The Chinese Institute of Atomic Energy was set up to train nuclear scientists. It is now the main research centre on nuclear weapons.

Overall administrative control of the nuclear weapons programme was given to Nie Rongzhen, a top party official and military commander. Nie was a veteran of the Long March and had been among the first Communist commanders to enter Peking. He is a friend and classmate of Deng Xiaoping. He brought the same vigour to the organisation of a nuclear weapons programme that, twenty years earlier, he displayed as leader of a band of guerrillas operating behind the Japanese lines. Nie had emphasised the importance of Chinese self-sufficiency in nuclear technology from an early time: 'We should and absolutely can master, in not too long a time, the newest technology concerning atomic energy in all fields. . . . There are people who think that as long as we receive assistance from fraternal countries, the Soviet Union and others, there is no need for us to carry out more complicated research ourselves. This way of thinking is wrong.' When the split with the Soviet Union came, a breakdown in co-operation over nuclear weapons – notably Khrushchev's refusal to hand a sample atomic bomb over to the Chinese – was the immediate cause of the breakdown in relations. China soon developed its own gaseous-diffusion plant to produce enriched uranium at Lanzhou in Gansu and exploded the first bomb at Lop Nor in 1964.

Wang Ganchang went on to become not only one of the main contributors to scientific research in nuclear weapons but also a Vice-Minister in the Second Ministry of Machine-Building, which is responsible for both civil and military nuclear programmes. In 1984 he was head of a controlled nuclear fusion group, and a member of the Standing Committee of the National People's Congress.

In 1984 Nie Rongzhen, by then aged 85, was a member of the Politburo, the CCP Central Committee, Vice-Chairman of the Military Council of the Central Committee and Vice-Chairman of the Standing Committee of the National People's Congress.

*   *   *

When two underwater explosions blew up the Greenpeace ship *Rainbow Warrior* in Auckland Harbour in 1985, killing one of its crew, the world was shocked. However, David McTaggart, the Chairman of Greenpeace, was not totally surprised. He suspected the French government, and he had reason to know the lengths to which it would go to protect its nuclear testing programme. When, eventually, the details of the story came out, it turned out that the man who had suggested that action be taken to prevent the *Rainbow Warrior's* voyage of protest to Mururoa was Admiral Henri Fages, head of the nuclear testing centre. He held a meeting with the then Defence Minister, Charles Hernu. Hernu proceeded to order the French intelligence service to 'forecast and anticipate the actions of Greenpeace'.

The French nuclear testing centre as well as the nuclear weapons laboratories are under the control of the Commissariat a l'Energie Atomique, the CEA. The CEA was set up in 1945 with considerable autonomy over its own affairs. It has been described as a 'state within a state'. At the early stages of the French nuclear weapons programme, during the unstable period of the Fourth Republic, successive politicians were either opposed to the acquisition of nuclear weapons or ambivalent about it. All the drive came from the élite group of engineers and technocrats in the CEA. The leader of this group, and the director of the CEA for seven years in the 1950s, was Pierre Guillaumat, son of a First World War general and a graduate of the Ecole Polytechnique. It was Guillaumat himself who took the decision to set up the nuclear weapons programme in the CEA (under a misleadingly named 'Bureau of General Studies'), following a meeting of politicians, military leaders and nuclear experts on Boxing Day of 1954, at which the political and military leaders were unable to come to an agreed decision. For several years, knowledge of the existence of the nuclear weapons programme was kept secret even from highly-placed officials in the CEA and many Cabinet ministers. A number of French scientists opposed the weapons programme, and all the impetus came from the engineers, who held the top jobs in the organisation.

Today the CEA's military activities are under the control of a separate department within the CEA, the Directorate of Military Applications (DAM). This body now employs between 6500 and 10 000 people, and carries out both fundamental research into nuclear weapons and the industrial production of nuclear warheads. Its activities are conducted at six 'Study Centres' throughout France (in Bruyeres-le-Chatel, Limeil, Aquitaine, Ripault, Valduc and Vaujours). The centre at Ripault carries out research into implosion and detonation and is likely to have been a centre for research on the neutron warhead.

The staff of these Study Centres are mostly Armament Engineers (Ingénieurs Généraux de l'Armament), trained in the prestigious 'Grandes Ecoles' of France. These colleges are distinct from the ordinary universities. They cream off the élite of French students and dominate entry into the highest posts of French government. More than 90 per cent of those who hold leading positions in the development and production of nuclear weapons are educated in the Ecole Polytechnique, the most prestigious of the Grand Ecoles. They are called 'les X', after the crossed cannons which form the emblem of the school. It has been the engineers of this élite academy who have provided the driving force of France's nuclear weapons programme.

Not only is the CEA's budget largely under its own control, but even its choice of military research and development is substantially independent. General Charles Ailleret, director of the first French atomic bomb programme under General de Gaulle, said when the

Directorate of Military Applications was set up in 1957, that this new body represented 'what we (the military) had always tried to avoid, (a situation in which) the military had only a quite remote control over the ways in which funds transferred from the defence sphere to the CEA would be used'.

When and how the neutron bomb programme began remains a secret. What is quite clear is that its development preceded the setting of a military requirement. The CEA's annual report in 1973 referred to the development of a neutron bomb and warheads for tactical missiles, while according to the government's Defence White Paper, the doctrine for French nuclear weapons remained 'dissuasion of the strong by the weak' – the use of large-yield nuclear weapons to deter attack. With its enhanced radiation characteristics and lower blast, it was by no means clear how the new weapon would fit into the official strategy. At present, the French government has adopted a strategy based on the use of tactical nuclear weapons outside French borders as a 'final warning', but there is no clear role in this strategy for the neutron bomb. Indeed, the French government has not yet announced that the bomb is deployed, although its development is essentially complete.

The approach of the French engineers and scientists in this instance was similar to that advocated by Edward Teller: 'It is preferable not to ask military people what they want, but rather to push scientific research to its limits. Military needs will follow'.

*   *   *

In all countries, the concentration of expertise in nuclear weapons in nuclear weapons laboratories has given their spokesmen notable influence and authority. In the USA, not only are their views on the development of new warheads listened to with respect, but they also advise generally on new weapons developments and on strategic and defence policy. Herbert York, the first director of Defense Research and Engineering at the Pentagon, describes how this process began during the Second World War: 'Not only did they invent and build new weapons, they actively promoted them and participated in deciding when and even how these new instruments should be used. Scientists became deeply embedded in the policy process during the war period because the technologies they developed in effect redefined the very nature of warfare and national security.' In Britain, the Director of

Aldermaston has similarly been involved in policy discussions over the future of the British independent nuclear deterrent; in France, the CEA has lobbied for the adoption of the neutron bomb.

In the West, the nuclear scientists of the weapons laboratories have particularly used their influence to oppose test bans. Teller spoke strongly even against the Limited Test Ban Treaty, which prohibited the testing of nuclear weapons in the atmosphere. He lost this battle, although as a sop to the laboratories, the Atomic Energy Commission agreed to conduct a test programme at an even faster rate than before.

The Comprehensive Test Ban Treaty has received adamant opposition from the laboratories. President Carter came into office committed to immediate achievement of a full comprehensive test ban. Considerable progress had been made in 1978 towards a treaty, in the negotiations at Geneva. But in the summer of 1978, the Energy Secretary and former Chairman of the Atomic Energy Commission, James Schlesinger, approached Carter with the directors of the Los Alamos and Livermore laboratories (Harold Agnew and Roger Batzel). They based their arguments on the need to develop and test the coming generation of nuclear warheads for more accurate missiles. They attacked the Treaty before the House Armed Services Committee and physicists at the laboratories sent a barrage of critical letters to key members of Congress. Harold Agnew later said: 'I met with President Carter for almost two hours on the (Comprehensive) Test Ban Treaty, through Schlesinger's intervention, together with Livermore's Roger Batzel. We influenced Carter with facts so that he did not introduce the CTB, which we subsequently learned he had planned to do. There's no question in my mind that Roger and I turned Carter round . . .'. Carter was persuaded to narrow and harden the American position, dashing hopes of a Geneva agreement.

In Britain, the scientists at Aldermaston and senior nuclear administrators in the Ministry of Defence have similarly opposed a Comprehensive Test Ban Treaty. Their views have been an important factor in the government's reluctance to agree to a Treaty. In 1980, Dennis Fakley, the Assistant Chief Scientific Adviser (Nuclear) in the Ministry of Defence told a House of Commons Select Committee: 'I had the dubious pleasure . . . of actually being a member of our team negotiating for a comprehensive test ban treaty for 18 months.' He went on: 'Under all the options that we consider to be realistic we still felt we would be able to satisfy the Trident requirements.'

In France, the loud protests of the physicists in the CEA, when the Socialists temporarily suspended nuclear tests in Mururoa, had such effect that three days later the tests were resumed. France has consistently opposed the Comprehensive Test Ban Treaty.

Only in the Soviet Union – where, although a powerful sub-élite, the scientists are dominated by the political leadership – has it been possible for a unilateral moratorium on tests to be undertaken and for comprehensive test bans to be endorsed.

Are the laboratories under political control? It seems likely that incremental work to develop and modify existing systems is controlled and controllable by the ministries of defence. But the startling innovations, the new ideas for weapons that no one had asked for because they had not been conceived, come from the laboratories themselves. In the Soviet Union and China, where control over the development of nuclear weapons is tightly held by the political leadership, incremental development appears to be preferred. In the USA, in contrast, the diffuse nature of control over the development of nuclear weapons and the competitive nature of the laboratories and the scientists acts as an engine for qualitative innovation. For so long as the political leadership accepts a need to maintain and encourage technical innovation in the weapons laboratories, the direction that new weapons will take – and hence the degree of instability in international military competition – will remain fundamentally unpredictable and uncontrolled.

3 The Defence Contractors

[The missile plants – the US aerospace corporations – the Soviet Design Bureaux – Qian Xuesen and Chinese rocketry – British Aerospace – Aerospatiale – technological momentum and the role of the defence contractors]

The final assembly hall in the General Dynamics Convair Division plant at San Diego is a brilliantly-lit, air-conditioned hall. Blue signs hang from the roof, indicating different assembly stations. In each of these lies a partially-assembled missile, painted bright yellow. The missiles are slung in wheeled conveyors, on which they can be moved along the highly-polished floor. One technician attends each missile. Dressed in white or grey coats, with open-necked shirts, they peer at the equipment tightly packed in cylindrical sections. Making small adjustments, they bring the sections together.

About 6000 miles away, on the other side of the North Pole, similar technicians work on similar missiles in another plant. Above them a 'No Smoking' sign in Cyrillic characters is prominent. In another part of the factory, rocket engines are being tested, their nozzles emitting a fiery blast in an ear-splitting crescendo of noise. Elsewhere, programmers are huddled over computer consoles, their faces pale green in the reflection of the screens. As they tap the keyboards, streams of figures and diagrams are displayed.

It is here, in the laboratories and production plants of the nuclear weapons states, that the facilities are maintained and the techniques honed to design and build rockets and missiles capable of being automatically launched, flying halfway round the world, and delivering their warheads with an accuracy of less than 100 metres.

Those who make the delivery systems for nuclear weapons play an important part in deciding which weapons systems are built. But the nuclear states organize their research, development and production facilities in very different ways. This leads to important differences in the role of the missile-makers in decision-making.

*   *   *

Every year, in recent years, several hundred Ground-Launched and Sea-Launched Cruise Missiles have rolled off the assembly lines of

29

General Dynamics' Convair Division. Further north up the western
seaboard, 1547 air launched cruise missiles had passed through the
Boeing production lines at Seattle by 1983. Every year, the corporations seek fresh contracts to keep their plants in operation and their
businesses profitable.

Boeing and General Dynamics are the two largest defence contractors in the United States. With six other corporations – Lockheed,
Northrop, Martin Marietta, Rockwell International, McDonnell
Douglas and General Electric, they account for over 80 per cent of the
prime contracts for nuclear delivery systems.

In the 1980s, business is good for Boeing and General Dynamics.
But it has not always been so. In the 1960s, Boeing had to lay off 56 000
workers, after losing a series of contracts, including those for a Navy
surface missile, an Air Force strategic bomber and a supersonic
transport plane. General Dynamics made losses on several contracts
for jets and missiles and was said to be 'a prime candidate for
extinction'. The corporations operate in a highly competitive market
in which profits and personal success within the company often depend
on winning government contracts.

Some miles from the Convair factory, in a plush downtown office,
Vice-President Leonard F. Buchanan sits at his desk. On the wall are
photographs of the weapons and missiles which have passed through
Convair's plants into military service. On his desk are thick reports,
and a handful of urgent memoranda relating to the plant's newest
weapon system.

Buchanan is well aware of the key importance to his company of
decisions made at the earliest stages of a weapon's life-cycle. In order
to bid successfully for a contract, it is necessary to have developed the
technology and skills in the area concerned up to the current state-of-
the-art. If the company has actually conceived of the weapon system,
its chances of pursuing R & D will be so much the better; if it has
undertaken R & D work, it is more likely to be awarded the contract
for development work and production – the stage at which the real
profits are made.

The contractors, therefore, pursue a good deal of R & D work
independently, using their own funds – though they can reclaim funds
from the government to support this work as well as the costs of
submitting bids and proposals. They maintain advanced teams with
specialities in rocket engineering, electronics, jet propulsion, and
guidance systems, and they sub-contract specialised work to thousands
of smaller companies. These teams seek out concepts for new weapons

before any political or military need for them has been articulated.

Formally, the process is that the Armed Service identifies a need for a new weapon and gives contracts to industry after they have been approved by the Department of Defense out of budgets voted through by Congress. If this was how the system worked in practice, it could be said to be under political control. In fact, the contractors dominate the development of the technologies on which the new weapons are based, and actively influence the shaping of the military requirement.

'We recognise it is the Government agency that must prepare the "Mission Element Need Statement",' said the director of corporate planning at Lockheed, 'but we feel that industry may be able to provide valuable inputs to the agency as it defines its mission needs. In our company, we try our hand at drafting the Mission Element Need Statement ourselves.' An official of North American puts the same view even more bluntly: 'Your ultimate goal is actually to write the RFP (the programme requirement), and this happens more often than you might think.'

In order to anticipate military requirements, the contractors maintain their own operations analysis divisions, which undertake strategic and economic studies. They also take on senior personnel from the military and the Department of Defense, in considerable numbers. The Convair Division's Vice-President for research and engineering was formerly Director of Air Warfare and Tactical Missiles for the Air Force. A former Vice-President and General Manager of the division was Assistant Secretary of the Air Force for Research and Development.

The companies obtain early access to intelligence assessments of Soviet weapons (on which projections of the need for new US weapons are formally based); sometimes they are even invited to prepare parts of these intelligence assessments themselves.

Most important, however, is their technical edge. In the United States, new weapons systems sell on quality and technical sophistication. Recruiting top technologists and maintaining a lead in relevant fields is the key to success in obtaining conracts. Given the competition for contracts among the aerospace giants, this helps to explain the restless innovation and rapid generation of new weapons types by the US defence industry.

The major innovations in the Cruise missile, other than its tiny warhead, were its highly efficient turbofan engine and its TERCOM navigation system. TERCOM had originally been patented in 1958. It was associated with an early missile programme which was cancelled in

1963, but the contractor, E-Systems, kept it alive using company funds. Later developments in microelectronics and microprocessors made possible the semi-intelligent, terrain-following, course-changing capability of the Cruise missile. The turbofan engine was developed by the Williams Research Corporation of Michigan, as a part of work for another earlier programme, the Short-Range Attack Missile. These separate lines of innovation were combined to bring about Cruise.

The synthesis of these elements appears to have come about – unusually – in the Pentagon itself. Civilians in the office of the Director of Defense Research and Engineering saw the possibilities of integrating these components to produce a long-range, strategic Cruise missile, longer in range than both the short-range bomber-launched missile the Air Force was investigating, and the short-range anti-ship missiles the navy wanted. The companies were then invited to bid for the development contracts. General Dynamics won the contract for the naval missile. There was a competitive fly-off for the air-launched missile between Boeing and General Dynamics, which Boeing won.

Once a new missile is under development, the company's interest in proceeding to production is very strong. Therefore enormous efforts are made to lobby for the weapon, both within the Services, in the Pentagon, and with Congress.

The companies maintain offices in Washington to keep in touch with government thinking and influence the government. They lobby Congress and make large donations to the political committees of key congressmen and senators. This process of patronage plays an important part in American politics, and influences government decisions both to place and to cancel military projects. Senator Henry 'Scoop' Jackson is known as the 'Senator for Boeing', and in his powerful position as Chairman of the Senate Armed Forces Committee, he consistently argued for the interests of the defence contractors, particularly Boeing. It was Jackson who recommended to President Nixon the appointment of James Schlesinger, whose doctrine of 'countervailing force' and support for a new generation of accurate counterforce weapons has done much to shape the current generation of new American nuclear weapons.

While a weapon is in the development stage, the company keeps in close touch with the Armed Service which has ordered it. Normally a project office is set up, for example at an Air Force base, and it is common for technical people from the Air Force and the company to be exchanged. The project office and the contractor have a common interest in the project's success, and help to form a powerful lobby for the weapon's adoption.

The contractors exert influence at higher levels of the military service and the Department of Defense through Scientific Advisory Boards, such as the Defense Science Board which advises the Secretary of Defense. Normally several senior officials from the defence corporations sit on these boards.

It is also common for executives from the defence industries to take jobs in the military services and in the Department of Defense.

In this way, a 'defence community', with shared assumptions on the need for new weaponry, has grown between the contractors, the Services, and the Pentagon. While contractors, services and Pentagon are often at odds, there are also strong shared interests in the continuing development of new weapons.

*   *   *

Early in the 18th century, Peter the Great built up the first Russian industry in order to manufacture arms for his army. His Soviet successors have continued to give the defence industry pride of place in their industrial planning. In the 1930s a central feature of Stalin's rapid industrialisation programme was the creation of a powerful defence industry, since a military assault was expected from the capitalist powers. Today the defence sector continues to be given the highest priority. Of all qualified engineers and scientists 50–60 per cent work in defence, and defence workers are paid higher than average salaries. The sector can even commandeer resources from the civilian economy.

Like the rest of the economy, defence is nationalised and centrally planned. The defence industry is controlled by no less than nine Defence Industry Ministries. The Ministry in charge of strategic missiles is the Ministry of General Machine-Building (Minobsche-mash). It has prime responsibility for the development and production of delivery vehicles for strategic weapons. Other ministries, such as the Ministries of Radio and Electronics, supply components.

Associated with the Ministry of General Machine-Building are a number of Design Bureaux, each with responsibility for designing a particular line of missiles. These are named after the leading rocket designers who built them up. These men are now seen as legendary figures of superhuman energy, who overcame the immense technical and practical problems of building intercontinental missiles in an economy that was still backward in comparison with the Western states. For example, there was Korolev, who built the first Soviet ICBM. Khrushchev wrote: 'we had absolute confidence in Comrade Korolev. When he expounded or defended his ideas, you could see passion burning in his eyes, and his reports were always models of clarity. He had unlimited energy and determination, and he was a brilliant organiser'. The Korolev Design Bureau developed the SS-1, SS-2, SS-3, SS-6 and SS-8: 'one rocket is being tested, the next modification is on the drawing board, while the third is being conceived'. In 1954, a deputy of Korolev's, Yangel was given a Design Bureau of his own. This designed the SS-4 and, it is thought, the 'heavy' SS-7, SS-9 and SS-18. Another Design Bureau was formed under Cholomei, a designer of naval missiles; this produced the SS-11, SS-17 and SS-19. Another was formed under Nadiradze to develop solid-fuelled rockets. It was this bureau which designed the SS-20.

Aleksander Davidovich Nadiradze was born in 1914 and graduated from the Moscow Aviation Institute in 1940. During the war he worked as a designer on the 'Katyusha' rocket artillery system. This was mounted on the back of a truck and was capable of firing a salvo of 48 rockets in rapid succession; it was used against German troop positions as the Russians fought back through western Russia and Poland. In 1958 Nadiradze was entrusted with the development of solid-fuelled missiles, perhaps in response to the US Minuteman programme. The result was the SS-13, test-flown in 1965. Nadiradze was awarded a Lenin Prize in 1966.

As the Bureau was asked to design solid-fuelled rockets of longer range, technical difficulties arose. Complex problems of fuel chemistry, materials science and in-flight control presented themselves, and these were not satisfactorily overcome for many years. Although the SS-13 entered service, only sixty were deployed. Its successors, the SS-14 and SS-15, did not reach full-scale production, and the third stage of the long-range SS-16 also proved unreliable. The SS-20 incorporated the first two stages of the SS-16 and was then deployed in large numbers as the first successful solid-fuelled missile, though its range was less than intercontinental.

The Design Bureaux are responsible for design and development work on strategic missiles. They are not responsible for either basic research or production. These are carried out by other bodies. Research is undertaken by the Rocket Research Institute and other institutes connected to the Ministry of General Machine-Building. Production takes place in more than twenty plants run by the Ministry's four Production Administrations. Normally the Design

Bureau develops a working prototype before a system is handed on to the Production Administrations for manufacture.

In general, the Soviet Union has lagged behind the USA in the branches of military R & D connected with missiles. The USA was the first to develop MIRVs, solid-fuelled rockets, super-hardened silos, submarine-launched missiles and long-range cruise missiles. Admiral Turner, a director of the CIA, told a Congressional committee in 1977, 'an overall assessment would be that we are well ahead of them in military technology. With brute force techniques, however, they do achieve about the same end result in many areas that we do with much more sophisticated techniques'.

The Soviet designers are certainly capable of innovation when it is required – witness their development in the 1980s of a new generation of accurate cold-launched, solid-fuelled 'silo-busting' ICBMs. However, most of their innovations have been brought about in response to US initiatives, and at the direction of the Politburo, the Defence Council and the armed services.

Why is missile development in the Soviet Union less innovative than in the United States? One factor is that the Design Bureaux lack the sharpness of competition which stems from the struggle of independent corporations for contracts. Although there is some deliberately fostered competition between Design Bureaux – sometimes two bureaux have been set to design the same type of missile – the Design Bureaux are internal departments of the Ministry of General Machine-Building, and their instructions come from above. They are not independent decision-making bodies, as are the US corporations.

A second difference lies in the way in which applied research is organised and funded. In the United States, the defence contractors undertake their own research to stay abreast of the competition, and the companies have access to both internal and government funding for advanced development and research programmes. They can commission research and sub-contract to other companies at their discretion, and they continue to carry out research to incorporate improvements in a system after the initial development contract has been won. The system gives those directly involved with current projects the power to push forward technologies appropriate to their needs and systematically rewards technical innovation. In the Soviet Union, in contrast, research and development are institutionally separate, and separately funded. There is no particular incentive for the designers to push their designs forward beyond the 'state of the art' prevailing at the time they drew up the specification, and indeed they prefer to avoid complications and the risk of failure. Nor do the research institutes have an incentive to press new technology on the designers, since the funding of research does not depend on its results being incorporated into development projects.

A third difference, linked to the second, is the distinction between design work and production. The Design Bureaux do all the development work up to the point a missile is ready for series production. They do not actually carry out the production themselves. In the United States, the corporations have a strong incentive to develop a system to the highest state of technical sophistication possible, since this makes it more likely to be accepted for the production run, in which the real profits are to be made. In the Soviet Union this incentive is lacking.

A fourth difference lies in the method for making decisions on the development of new weapons in the early stages. In the United States, there is considerable power of initiative at a low level, and the corporations and the Service project offices can nurture new ideas for some time on their own. In the Soviet system, in contrast, even relatively minor decisions are considered in detail at several levels of the bureaucracies of both the military and the defence industry. More people in higher positions have to approve. Consequently the decision-making tends to be more conservative.

In the Soviet Union, lines of demarcation between the military services and the defence industry are sharp. As a rule people do not move from one to another, and the strong vested interests that develop around particular weapons systems between corporations and military in the United States have no parallel. Indeed, there has been antagonism between the defence industry and the military, with the military sometimes critical of the industry's capability to provide weapons of the quality and type it requires at the time it requires them. Equally, officials of the defence industry have been irritated by the military's insensitivity to problems of technology and supply.

This controversy surfaced in a public row in 1984 when Marshal Nikolai Ogarkov, then Chief of the General Staff, said in an interview with *Red Star* that 'weapons based on new physical principles' were required to keep up with the Americans, and implied that the defence industry was incapable of keeping up. Dmitri Ustinov, the Defence Minister, replied testily that 'the defence industry is capable of producing any weapon'.

The military may complain, but the defence industry ministries are in a dominating position. Not only do they have a monopoly over weapons supply, they also have great influence in the Politburo.

Several leading Politburo members worked their way up through the defence industry ministries and heavy industry, not surprisingly, in view of the Soviet Communist Party's preoccupation with industrialisation and security. Brezhnev himself had a background in engineering and had served as Central Committee Secretary for military affairs, the defence industry and the rocket and space programmes. At the end of Brezhnev's life, the Politburo also included Dimitri Ustinov, who became Commissar of Armaments in 1941, Nikolai Tikhonov, a former steel industry administrator, Grigorii Romanov, from the shipbuilding industry and Shcherbitskii and Kunaev from the metallurgical industry. In 1985, a new five-year plan gave continued priority to heavy industry and the defence industry.

For the defence industry in the Soviet Union, as in other industries, the key target is to fulfil planned production norms. Factory managers and workers receive bonuses for successful fulfilment of plans on time. The industry therefore resists sophisticated, complicated weapons which are likely to be late, and prefers the manufacture of relatively simple, standard systems in bulk. These preferences show in the type, numbers and quality of nuclear weapons the Soviet Union deploys.

*   *   *

China was the first country to use rockets for military purposes, when the Mongol Emperor Kublai Khan used them on the battlefield in 1274. However, China was the last of the five major nuclear powers to possess an intercontinental missile.

The CSSX4, which trundled through the streets of Peking in the May 1st parade, is the fourth and most recently deployed Chinese land-based missile, with the longest range (7000 miles). The earlier CSS1 and CSS2 were medium-range missiles, capable of 700 miles and 3000 miles respectively. China has only a handful of intercontinental missiles, although there are over 100 nuclear missiles of shorter range and older design.

As in the Soviet Union, the plants responsible for building these missiles are departments of the Defence Industry Ministries.

Qian Xuesen is regarded as the 'father' of modern Chinese rocketry. He was born in Shanghai, the son of a businessman. When he was 21 he went to the United States and studied aeronautics and aerodynamics at the Massachussets Institute of Technology. He then went on to Caltech, where he worked with Professor Theodor von Karman, a leading scientist of aerodynamics, who interested him in jet propulsion. There Qian contributed the 'Qian formula' for rocket propulsion. Later he won his doctorate and was given charge of a laboratory working on research into supersonic flight. During World War II Qian became director of the Rocket Section of the US Defense Scientific Advisory Board and a colonel in the US Army. After Germany's defeat, he was head of a group of US scientists sent to dismantle the German rocket centre at Peenemunde.

Qian returned to China in 1947. There he married Jiang Ying, a teacher of singing and music. In the same year he went back to the United States, and became director of the Guggenheim Jet Propulsion Laboratory, and a Professor of Jet Propulsion at Caltech. In 1950, a year after the Communist Revolution, Qian decided to return to China, but the FBI prevented him from leaving, and arrested him for sending documents to China. After 15 days he was released on bail, and for the next five years he continued his research at Caltech. Finally he was allowed to leave, in a swap with nine Americans being held by China. With his wife, children, and a heavy crate of scientific papers, he was back in China in October 1955.

On his return, Qian was appointed director of the newly-established Institute of Mechanics of the Academy of Sciences. He joined the Communist Party in 1958 and became an alternate member of the Central Committee in 1969. Unlike many of his contemporaries, he avoided the purges accompanying the Cultural Revolution. In 1982, he was the vice-chairman of the Science and Technology Commission for National Defence, the top Chinese military R & D organisation. This body is responsible for overseeing the development of nuclear weapons and delivery systems and allocating military R & D funds, and reports directly to the top Military Commission.

The early Chinese missiles were based on Soviet models – for example, the CSS1, a liquid-fuelled missile, deployed either in the open or in caves, was built on the model of the Soviet SS-2. The later missiles owed more to the innovations of the Chinese scientists. The CSSX4 resembles the US Titan rocket.

The missiles are produced in plants belonging to the Seventh Ministry of Machine-Building (now the Ministry of Space Industry), one of eight defence industry ministries created along Soviet lines.

Like other Chinese ministries, this is vertically organised, and runs its own plants for producing components and basic materials. The Chinese defence industries run their own steel plants, for example, instead of purchasing requirements from civilian sectors.

Unlike its Soviet counterpart, the Chinese defence industry is not given a dominant priority in industrial development. Following the break with the Soviet Union, Mao Zedong rejected the Soviet model of building up heavy industry. Instead, agriculture and light industry was given priority. In subsequent Five Year Plans, defence production has consistently been given lower priority than the development of other sectors. When Deng Xiaoping announced his 'Four Modernisations', the modernisation of the military was given the lowest priority of the four. The Chinese armed forces are still equipped with arms mainly of Korean War vintage.

Nevertheless, special efforts were made to protect the development of nuclear weapons from the disruptions of successive waves of political turmoil. Many of these have involved the defence industry, directly or indirectly. For example, Mao used the PLA to support his purge of the Party in the Cultural Revolution, and the factional strife extended to the Seventh Ministry as well as other parts of the party and government. The dispute between Mao and Lin Biao must have involved the defence ministries, since Lin was minister of defence industry before his mysterious death in a plane crash near the Soviet border. Deng Xiaoping dismissed the heads of all the defence ministries in 1978. At the same time, he reduced defence expenditure and cut the size of the PLA by half. In these upheavals and conflicts, the development of modern missiles, in a basically agricultural country with a low per capita income, must have been a particularly difficult enterprise. Nevertheless, the Politburo has been determined to pursue the nuclear programme irrespective of political and economic obstacles.

*   *   *

The British and French defence industries fall somewhere between the pattern of the US industry, with its large privately owned corporations, and the Soviet and Chinese state defence industry ministries. In both countries defence contractors play an independent role, but their influence in decision-making is much less than in the USA. Both countries have corporations which play an important part in developing nuclear delivery systems: the recently privatised British Aerospace in Britain and the nationalised Aerospatiale in France.

British Aerospace is the prime defence contractor in Britain. It has a turnover of about £2500 million, of which sales to the Ministry of

Defence accounts for almost 40 per cent. It is also a major arms
exporter, with defence accounting for almost 80 per cent of turnover
and for all of its profit.

Its largest section, the BAe Aircraft Group, designs and produces
both civil and military aircraft, including the nuclear-capable Tornado,
the Harrier, the Sea Harrier, the Jaguar and the Nimrod. It also
subcontracts on the McDonnell Douglas Phantom and the General
Dynamics F-111 and F-16. The second major section is the BAe
Dynamics Group, which manufactures guided weapons systems and
missiles. It also builds satellites and space systems. It was this group
which acted as principal contractor on the Chevaline project.

The Chairman of British Aerospace is Sir Austin Pearce. As a child,
Pearce was caught in a German air raid. The feeling of terror and
helplessness he experienced left a strong impression. He visited Nazi
Germany before the war and he regarded Chamberlain's appeasement
policy with disgust. Later he spoke to people who had been in Japanese
POW camps and heard what had been done to them. These
experiences made him determined that Britain should never be
defenceless again.

Admiral Sir Raymond Lygo, who was previously head of the
Dynamics division, is the Managing Director of British Aerospace. He
left school at 14 and became a messenger on *The Times*. Then he
joined the Fleet Air Arm, rising to take command of the *Ark Royal*
and to become Vice-Chief of Naval Staff. Then he left the Forces to
join British Aerospace. Lygo reflects a growing trend – in 1984, 680
senior serving officers and civil servants left the Ministry of Defence to
join the defence industry.

Lygo has adopted some of the defence industry's dislike of the hand
that feeds it. Of Chevaline he said: 'It was run in a way from the
Ministry of Defence that proved to be totally ineffective and in the end
the only options left were to bring in the professionals, industry, to run
this programme and it just so happened that British Aerospace was
brought in to manage the programme and from that moment on once
they had sorted it out it ran to cost (and) time, but (until then) it hadn't
been managed.'

British Aerospace took over the management of the Chevaline
project in 1977. Until that time it had been in the hands of the nuclear
scientists. The project had been delayed due to uncertainty among the
politicians and in the MoD as to whether it should proceed, and
because of the real difficulties involved in developing multiple re-entry
vehicle technology, which had not been acquired in Britain before.

Indeed, after the cancellation of 'Blue Streak', Britain has retained no indigenous capacity to design and manufacture strategic missiles. It is the only one of the five major nuclear powers to lack this capacity. All Britain's missiles are bought from the United States.

Partly for this reason, the defence contractors do not play an influential role in decisions concerning the independent nuclear forces in Britain. In the case of Chevaline, it appears that British Aerospace was brought in to act as a project manager. It is unclear whether it played any role in decision-making. On the Trident decision, British Aerospace was not consulted at all. Lygo called this omission 'extraordinary'.

In fact, US contractors play a more important role in decisions about the independent British nuclear deterrent than do British contractors. Decisions taken by US contractors to close down the Polaris production line influenced the government's decision to embark on Chevaline. US contractors are now doing studies for the Ministry of Defence on the choice of dimensions for the British Trident.

*   *   *

Unlike Britain, France makes its own missiles. Its defence contractors have manufactured all the delivery vehicles for French nuclear weapons, including the medium-range land-based ballistic missiles, based at the Plateau d'Albion in the southeast of the country, the Mirage IV nuclear bombers, the nuclear missile submarines, and the 'Pluton' short-range tactical nuclear missiles. Several of these systems are being modernised and replaced.

The industrial base required to maintain this arsenal employs about a fifth of the industrial work-force and swallows a third of national expenditure on research and development. Among a number of important defence contractors, Aerospatiale is the most important and acts as the prime contractor for all France's missiles, from Exocets to nuclear missiles with a range of several thousand miles. The links between Aerospatiale and the government are extremely close. The company's president, until 1983, was Jacques Mitterand, the brother of President Mitterand and a former commander of the strategic forces.

The French defence contractors are much more closely linked to the government than their US counterparts, though they are not parts of the government like the Soviet Design Bureaux. Most of the large contractors are either nationalised or part-owned by the government. The government maintains a 'poste de contrôle' in every company, staffed by an engineer from the Directorate General for Armaments (DGA), the procurement arm of the Ministry of Defence. As in the Soviet Union, where the military monitors the defence industry, the purpose is to maintain quality control. There is a constant flow of personnel, especially Ingénieurs Généraux de l'Armament (armament engineers trained in the Ecole Polytechnique and the School of Armaments), between the DGA and the contractors. This highly-trained group of engineers has become a coherent élite within the industrial and governmental wings of French weapons procurement and it maintains the close ties between them.

Research and development are undertaken both within the defence companies and in government agencies. Sometimes concepts for new weapons are developed by the companies and pass up through the DGA to the government; sometimes they are formulated within the government and pass down. Within the government, military R & D policy is formulated by the Council for Defence Research and Studies (CRED) and the powerful top-level 'Groupe S' (Group for Planning and Strategic Studies). France cannot afford to develop a large number of major weapons projects and once a commitment is made to proceed with research and development, a project is unlikely to be cancelled. Decisions at the early stage are thus of crucial importance. These decisions are taken by administrators within 'Groupe S' and CRED on the basis of advice, information and opinion in the industrial and governmental arms procurement communities.

*   *   *

In all five nuclear states, the defence contractors form an important lobby for new weapons. In the Soviet Union no less than the United States, laboratories and factories need new weapons to keep their research and technological standards up to date and to maintain funds and morale. There are, however, very important institutional differences.

In the United States, the defence contractors are autonomous corporations, with a strong profit motive in getting new weapons approved. They carry out their own research and compete on technological quality. There is a close community of weapons experts which moves between defence corporations, military posts and jobs in the Pentagon, and has a dominating effect on the decision-making process. Effectively, the same community makes the new weapons and decides which new weapons should be made. The key moment of decision for most weapons is the stage before research and development funds are committed, and it is then that opinion in this community, usually ahead of the public debate, is most critical. The power of the defence contractors means that initiatives for new weapons 'float up' from the research facilities operated by the contractors and the military.

In France, too, there is a similar community between the Directorate General for Armaments and the nationalised defence industries, based on graduation from the élite Grand Ecoles. It is less interlocked with the military than in the United States, and decisions on new systems are taken in a centralised way. Yet, as in the United States, there is a close community between those responsible for developing the weapons and those who select them.

Such a community does not exist to the same extent in Britain, where there is a gulf between the thinking and mannerisms of the mandarins of the Ministry of Defence and the industrialists of the defence industry.

In both the Soviet Union and China the defence industries are a part of government, under much greater political control than in the West. Research and development work is institutionally separate from production, and there is more compartmentalisation between military, defence and government bodies. Even at the early stages, weapons projects are closely and bureaucratically controlled. The defence industry itself has considerable influence in the top ranks of the Party, but the Party, not the defence industry, is in control.

Insofar as technological momentum drives the arms race, the existence of autonomous, competing, defence contractors, developing their own proposals for new weapons, is a key source of instability. Insofar as political control over defence contractors is necessary to bring the arms race under restraint, this control exists now in the Soviet Union and China. In the western states, especially the United States it is inadequate.

4 The Military

[From strategic bombing to nuclear commands – the role of the Strategic Air Command and the US Services in nuclear weapons procurement – the influence of the Soviet military – Admiral Fieldhouse and the Service Chiefs in the Ministry of Defence – the French military – the People's Liberation Army – the role of the military in nuclear weapons decisions and the influence of military thinking]

From the air, the city was completely shattered. The streets stood out, white and empty, from the blocks where tall buildings had once stood. Now they were gaunt shells, their roofs gone, their floors gutted. Windows looked eerily out from rooms which had disappeared. In places the internal walls remained, exposed, showing, as if in a historical relic, the intricate pattern of the city's interrupted life. Some buildings had miraculously escaped damage, others were only lightly charred. Elsewhere, they had been completely reduced to rubble, or transformed into unsupported pillars of brickwork and mortar. Block after block, street after street, the view was the same. A once great city lay in ruins.

The city was Berlin in 1945. The destruction was visited upon it by the Strategic Bombing Offensive of the Western Allies and by artillery fire from the Soviet ground forces.

Within a few years, the men who commanded the strategic bombing and artillery attacks of the Second World War were to be placed in charge of nuclear weapons. Swollen by war, the military establishments of the major powers had reached, by 1945, a position of great influence in the affairs of the State, both in the West and in the East. The development of weapons that could obliterate whole cities at a stroke put potentially enormous power into the hands of the military, and at the same time dramatically altered the nature of war. It is not surprising that the military were to play a vital role in nuclear weapons decisions.

*   *   *

Curtis LeMay, the chubby, cigar-smoking general who had commanded the fire-bombing of the Japanese cities, became the first commander of the Strategic Air Command (SAC). He presided over the early American build-up of a large force of nuclear bombers.

In 1957 LeMay was visited at his Omaha headquarters by Robert Sprague, a member of the Gaither Committee, a special commission set up by the National Security Council to examine the vulnerability of American nuclear bombers to a surprise attack. Sprague asked LeMay to order a mock alert to test the bombers' readiness. To his consternation, none of LeMay's bombers were capable of taking off in the time it would take a Soviet attacking force to reach SAC's aerodromes after they had crossed the DEW (Distant Early Warning) Line. LeMay was unperturbed. He had no intention of absorbing a Soviet attack and striking back only in retaliation. 'If I see the Russians are amassing their planes for an attack', said LeMay, 'I'm going to knock the shit out of them before they take off the ground.' 'But General LeMay', said Sprague, 'that's not national policy.' 'I don't care', said LeMay. 'It's my policy. That's what I'm going to do.'

The Strategic Air Command today is in control of two arms of the 'Triad' of US strategic forces (the nuclear bombers and the land-based missiles), while the Navy controls the third (the nuclear submarines). In addition, tactical nuclear weapons are integrated into the forces of every US military service, so each service, Air Force, Navy, Army and Marine Corps, controls nuclear weapons. The Strategic Air Command's Commander-in-Chief is the Director of the Joint Strategic Targets Planning Group, which sets the targets of US and NATO strategic nuclear forces. In the event of war, it would be the SAC headquarters underground, at Offutt airforce base in Omaha, Nebraska, that would launch the US bombers and land-based missiles, on the orders of the National Military Command Centre in Washington. The Strategic Air Command also has considerable influence over the selection of its own weapons.

Each military service in the USA is responsible for its own weapons procurement and budget. The Air Force, like the other services, not only submits its own budget and receives funding from the Treasury, it also purchases weapons and manages development programmes. Although the civilian officials in the Office of the Secretary of Defense formally have authority over the Services' procurement and budget plans, in practice each service exercises considerable independence.

Over 90 per cent of the Pentagon's annual budget, currently running at more than $250 billion, is spent by the Services themselves. The budgets are drawn up by the Services, and then have to be approved by the Secretary of Defense, the Office of Management and Budget, the President and the Congress. A separate body within the Department of Defense, the Defense Systems Acquisition Review Council (DSARC), controls the acquisition of weapons and expenditure of funds.

These formal processes establish civilian control in principle. In practice, the amounts of money are so huge that, given limited time available for adjusting the budgets, detailed monitoring of every item is impossible. Programmes costing less than $25 million are passed through the Department of Defense without scrutiny by the civilian officials, and the DSARC process concentrates on 'major weapons systems' costing at least $200 million in research and development or $1 billion in production.

At the early stages in a weapon system's life, therefore, before it has become expensive, the Services have effective control. For example, the Air Force started the development of the MX missile with the award of two development contracts, one for $19.6 million to the Aerojet Solid Propulsion Co., for 'the design and fabrication of one second-stage propulsion system that could be applicable to USAF's proposed MX advanced intercontinental ballistic missile', the other for $4.6 million to United Technologies to develop and demonstrate a 'high-deflection, low-torque, moveable nozzle for the MX booster stage'. These contracts were awarded before Congress had approved funding for the MX. The Air Force claimed that the contracts were legitimate because Congress had made available funds under other budget lines for continuing development of the Minuteman missiles.

While the Strategic Air Command is the customer for the land-based strategic missile and bomber forces, the Air Force System Command is responsible for the management of research and development and production programmes. The development of missiles is co-ordinated by its Ballistic Missiles Office at Norton Air Base, San Bernardino, California. It was this office of the Air Force which implemented the MIRVing of US missiles, and later managed the Minuteman and MX missile programmes. Like the weapons laboratories, the Ballistic Missiles Office needs a steady stream of new weapons to keep its technical teams together and to maintain purpose and momentum. The Office is formally responsible to the Systems Command and the Air Staff, but in practice it reports directly to the Undersecretary of Defense Research and Engineering in the Pentagon. However, the Strategic Air Command, the ultimate customer, also maintains a watchful eye. During the development of the MX programme, a team from the Strategic Air Command was deputed to Norton to attend all the key meetings, to ensure that the weapon fitted in with Strategic Air

Command's war-fighting priorities. In this manner, the Strategic Air Command had an effective veto over decisions on design specifications, even though it was not formally part of the decision-making chain of authority.

The Service departments which manage advanced technology projects depend on new projects for their prestige and funding. Consequently they have common interests with the defence contractors. Once a new weapons system has been started, a powerful lobby for its continuation begins to build up, both within the Service and outside it.

In order to win approval for a new weapon, it is necessary to secure the consent of various levels of the Service hierarchy, and usually that of the Scientific Advisory Boards which counsel each service on the merits of new weapons concepts. These Boards are composed of scientists, corporation presidents, representatives of the weapons laboratories, and others from the 'defense community'. The advice of strategists, consultants and 'think tanks' such as the RAND Corporation is also important.

Once a new weapon system has gained the support of the Service, it will then be put forward in the Service's budget request. By this stage, a lobby for the weapon has already formed and as further development, trials and modifications are carried out, this lobby will support further development, even if, as normal, the weapon costs more than original estimates and performs below specifications. The situation is very different in the Soviet Union, where the separation of functions between the military services and the defence industry ministries results in the military acting as quality controllers, able and ready to reject systems which do not conform to the original specifications.

As development proceeds, the costs of the weapon programme increase sharply, and the weapon begins to come into competition with other new systems. At this stage rivalry between the Services becomes intense, since it is generally not possible to fulfil all the Services' equipment requests within the defence budget. The battles for funds are fought out within the Pentagon and in the Congress. Changes continue to be made to the Budget until the last minute.

As the lobbying proceeds, the Military Services benefit from their close links with the defence contractors and congressmen representing the districts in which contracts create employment. The Services cultivate support among Senators, who also need to be able to get on well with the Services.

Although the Services do not always get their way, they are not easy to restrain. McNamara, Kennedy's Secretary of Defense, managed to rein back the Air Force, which wanted a vast increase of missiles to 10 000 and an ABM system. He held them to 1000 missiles and a nationwide ABM was defeated. But McNamara did approve the MIRVing of the missile warheads and reluctantly authorised a limited ABM system. President Carter came to office deeply suspicious of the unending demands for new weapons, and intent on cutting back the defence budget and cancelling the Air Force's expensive B-1 nuclear bomber. He did cancel the production of the B-1 bomber, but despite the cancellation, the Air Force kept research and development work going, and when Carter had gone, and Reagan came to power, the B-1 went ahead. In the case of the MX missile, the Air Force's dogged pressure enabled the programme to survive Congressional and Presidential scepticism and technical failure in the development programme. The missile had been designed for mobile basing, but no effective mobile basing scheme could be developed.

In principle, the 'missions' of each Service, which are supposed to determine the weapons it needs, are determined by the Joint Chiefs of Staff. The Joint Chiefs of Staff also begin the formal defence planning process with a formal statement of threat, based in part on intelligence assessments of Soviet weapons programmes. However, each Service maintains its own intelligence agency, and the views of these different agencies often vary. In the 1950s, for example, when the Air Force wanted more missiles, it used its own intelligence reports to warn the Congress of a 'missile gap' which the reports of the CIA and other agencies did not support. Senator John Kennedy took the theme up effectively in his Presidential campaign and, even though the Air Force reports subsequently turned out to be wrong, a rapid period of US missile-building ensued, which the Soviets later matched. More recently, Air Force intelligence estimates of Soviet anti-satellite programmes have been used to justify developing new Air Force programmes.

The US Armed Services have also maintained their own strategic doctrines, at times in conflict with those of other services and of the Secretary of Defense. Since the 1960s, the Air Force has persistently held to a 'counterforce' doctrine (that is, a strategy of targeting missile silos and other military installations), even when the Secretary of Defense proclaimed US policy to be 'mutual assured destruction' (that is, a strategy of targeting cities and population centres) and the Navy's policy was one of retaliation. This is made clear by testimony of the Air Force General, and later Chairman of the Joint Chiefs of Staff, David Jones, to the Senate Armed Services Committee in 1979:

Senator Tower: General Jones, what is your opinion of the theory of mutual assured destruction?

General Jones: I think it is a very dangerous strategy. It is not the strategy that we are implementing today within the military but it is a dangerous strategy . . .

Senator Tower: Your professional military judgement is that it is a dangerous strategy and it is not one that we should follow?

General Jones: I do not subscribe to the idea that we ever had it as our basic strategy. I have been involved with strategic forces since the early 1950s. We have always targeted military targets. There has been a lot of discussion . . . about different strategies. We followed orders, but basically, the strategy stayed the same in implementation of targeting.

Senator Tower: Unfortunately, I am not sure that your opinion was always shared by your civilian superiors.

General Jones: I agree there have been some, including some in government, who have felt that all that we require is a mutual assured destruction capability. I am separating that from our targeting instructions in the field.

The pursuit of this strategy has led the Services to develop many more weapons than would be required for a policy of minimal deterrence, and to continue to justify building new weapons to attack Soviet weapons, even beyond the point at which most of the Soviet population would be killed. It also meant that weapons of greater accuracy were required. The US Air Force, and its think-tank at RAND, took the lead in pressing for these. Although the technical innovations which made greater accuracy possible came from the weapons laboratories and the defence contractors, it was the Services, especially the Air Force, which promoted the development of counterforce weapons.

To a mind steeped in military training, counterforce targeting has an obvious appeal. Yet its logic leads to a first-strike posture, since there is no advantage in destroying empty silos. Indeed, a first-strike option is included within the Single Integrated Operational Plan (SIOP), the target list and war plan of the United States.

*   *   *

The Soviet Union has five armed services, each with some responsibility for nuclear weapons. The most important is the Strategic Rocket

Force, which controls the Soviet land-based, intercontinental and intermediate-range missiles. The Navy operates ballistic and cruise missile submarines and tactical nuclear weapons on surface vessels. The Air Force has control of the long-range bomber force and tactical, air-launched nuclear bombs and missiles. The Ground Forces control shorter-range nuclear missiles and nuclear artillery. The Air Defence Forces operate the ABM missiles deployed around Moscow, which carry high-yield nuclear explosives.

Each of the Services has a responsibility for procurement, with a Deputy Commander-in-Chief of the Service in charge. Their 'Technical Administration' departments issue requirements and specifications for new weapons, monitor research and development, carry out tests on prototypes and exercise quality control on weapons coming out of the production lines. The services also pay for the weapons, though the prices are fixed by GOSPLAN, the central economic planning department, and probably understate true costs. Representatives of the military Technical Administrations are assigned to the defence industry plants which are involved in production, to co-ordinate with their managers and ensure quality control.

The Soviet Ministry of Defence differs from its counterparts in the USA, Britain and France in having no civilian representation. Its officials are also military officers. Hence the voice of the Ministry is the unified voice of the military. The Commanders-in-Chief of the Military Services are also Deputy Ministers of Defence. The Minister of Defence in 1985 was Sergei Sokolov, a tank and army commander with experience of active service during the war on the Western and Karelian fronts. Sokolov is a Candidate member of the Politburo.

The Soviet General Staff, which is responsible to the Ministry of Defence, co-ordinates the various services, and has a much more important role than its US equivalent. Responsible for monitoring and developing strategic doctrine, and for the composition and equipment of the armed forces, the General Staff settles conflicts for resources between the competing services and is thought to oversee the direction of military R & D. Its Military Science Administration and Scientific and Technical Committee help to identify new weapon requirements. The General Staff operates the Military Intelligence Administration (GRU) which assesses external threats (in contrast the American Defense Intelligence Agency, though run by the military, is responsible to the Secretary of Defense). The General Staff is the major source of military information for the Defence Council, a top national security decision-making body, and for the Politburo. It has also played an important role in arms control, with officers of the General Staff acting as members of the Soviet negotiating team.

The Strategic Rocket Force, the most important of the military services, was set up in 1959, following a radical reappraisal by Khrushchev of the role of the armed forces. Previously, under Stalin, the missiles had been under the artillery department. Indeed, artillery-men still dominate the Strategic Rocket Force, and Soviet rockets were conceived originally as an extension of artillery in a possible war in Europe. In 1959 however, Khrushchev declared that intercontinen-tal rockets would decide the future of any war and the military services were reorganised accordingly.

Military doctrine in the Soviet Union appears to have been more consistent than in the United States, perhaps reflecting the greater integration of the military and political leadership. There has never been, so far as one can tell, any equivalent in the Soviet Union of the highly abstract strategic thinking that has characterised the views of RAND and the US strategic community. Nor is there a community of civilian strategists as there is in the United States. Soviet strategists do not appear to have accepted the US distinction between 'counterforce' and 'counter-city' strategies, nor to have debated and swung between them in the way the Pentagon has. The published evidence suggests that Soviet policy is to strike at enemy military forces, industrial centres, power centres, strategic communication centres and centres of command and control.

The Soviets never held quite the same conception of deterrence as their US counterparts. There is no precise equivalent in Russian to the English word 'deterrence' – the Soviets use the term 'sderzhivanie' which means 'keeping out' or 'restraining'. The emphasis has been on preventing a nuclear war and an attack on the Soviet Union.

This is conceived in a broad way, so that in peacetime it includes not only the building of new weapons, to maintain Soviet strength and rough parity with the United States, but also a foreign policy designed to reduce the risk of war.

In wartime, the policy would be to restrain an attack by military means. There is no doubt that there are also pre-emptive strike options in the Soviet war plan.

While the military have an influential voice in Soviet counsels, and their expertise is respected, the Politburo retains overall leadership. The Party has traditionally maintained a high regard for military power, but has also been concerned to keep the military under its own control. It is characteristic of the difference between the US and Soviet systems that, while in the United States the military control their own nuclear weapon stockpiles, in the Soviet Union these are controlled by the KGB. In the United States, two Air Force servicemen man each missile silo. In the Soviet silos there are four men – two from the Strategic Rocket Forces, and two from the KGB.

*   *   *

Leaving Faslane behind, the submarine's dark shape moves down the Clyde. Inside its narrow hull, at the stern of the ship, the controlled chain reaction in the pressurised water reactor produces a steady supply of electricity, which powers the submarine. In the forward section are the cylindrical tubes which stand vertically, running the full height of the hull, in which the missiles are stored. In the middle of the submarine, the bulk of the 150-strong crew go about their work in a cramped environment packed with dials, switches, buttons and equipment. The submarine is lit with white flurorescent lights by day and dim red lights by night. Most of the crew are in their 20s and 30s. They will remain submerged in this environment for 60 to 70 days. All are highly trained, disciplined, and practised, so that together with their boat they form an effective fighting machine.

These young men bear a heavy responsibility. On each departure, the commander knows that he could be called upon to launch the missiles on this patrol.

One man with experience of such responsibilities is Admiral Sir John Fieldhouse. Born in Leeds in 1928, the son of Sir Harold Fieldhouse, KBE, CB, John Fieldhouse joined the Royal Navy as a Cadet at the age of 13. He served as a Midshipman on several ships, and then at the age of 19 joined the submarine service as a lieutenant. He spent his next six years in submarines and took command of the first nuclear-powered submarine, *HMS Dreadnought.* Later he commanded the aircraft carrier *HMS Hermes.* In 1967 he was given command of the Polaris submarines at Faslane.

In the following years he continued to rise through the senior ranks of the Navy, becoming an Admiral in 1974, and Controller of the Navy in 1979. In this position he was the senior naval officer concerned with procurement at the time the Trident decision was made. In 1982 he won renown as the commander of the British Task Force to the Falklands. In 1985 he became Chief of the Defence Staff, the top post in the British military. In this position he would decide how Britain would fight if war broke out and – in consultation with the politicians – at what stage the nuclear button would be pressed.

Fieldhouse is a genial admiral with a ready smile and eyes that wink alternately at odd moments. He likes good food and wine, and enjoys Channel cruising in his 28-ft Seal yacht. His family home is in Hampshire, where he has a son and two daughters. His wife, Midge, speaks Russian, Spanish and French and enjoys giving impromptu piano recitals. According to a newspaper article, 'Midge has been crucial to his successful career. Fieldhouse and Midge form an extraordinary double act in a service noted for characters.'

Fieldhouse is committed to the nuclear deterrent, which he says has kept the peace since the Second World War. In his view, the deterrent must be seen to work and to be the best available. Asked what he would say if another government were to cancel Trident, Fieldhouse replied, 'I would say I believe this country requires a strategic deterrent. Not perhaps today so much as in the long-term future. I know we have a nuclear proliferation treaty to prevent other nations getting nuclear weapons but I haven't much faith that that in the end over a length of time will in fact be totally successful. There hasn't so far in the history of mankind been an absence of madmen occurring one way or another in the world. Whoever was responsible for laying those mines in the Gulf of Suez (in 1984 mines damaged thirty ships in the Gulf and the Red Sea) could well in 20 years time be responsible for threatening this country with nuclear weapons. I would like my children not to have to face such blackmail and it is only with the possession of a strategic deterrent such as Trident that they will be safe from such blackmail.'

Although there was widespread concern in the Services about the possible impact of the Trident purchase on government spending on the conventional forces, those at the top of the services concurred with a decision which many saw as unavoidable. Fieldhouse supported Trident, but was determined that its budgetary consequences should not be concentrated on the surface ships. 'I am absolutely in no doubt whatsoever that from the present choices it is the best buy to use a "Which"-type term. . . . It's an opportunity that unless we take now we certainly shan't repeat on the same basis with the United States. . . . It would be quite ridiculous in my view to start such a project of extreme complication as a national project in the United Kingdom to build only a few missiles. As far as the submarine is concerned I am convinced it is the best vehicle for the foreseeable future in which to hide our missiles system so that whatever else may happen should it ultimately ever be required it will be available when it is required.'

Asked whether the Services had enough 'decision-making clout', Fieldhouse replied that they had not. Many politicians, he says, have no grasp of the military situation, and a more concerted effort was needed to put across the military point of view. The military, he says, are always weak in times of peace. It is difficult to imagine a senior military officer in the USA or the USSR making such a remark.

In Britain, unlike the United States, the Services do not manage the procurement of their own weapons. This is carried out centrally by the Procurement Executive of the Ministry of Defence. Nor do the Services dominate the Ministry of Defence – unlike in the Soviet Union. In successive reorganisations of the Ministry the influence of the Service Chiefs has been gradually eroded. Against their wills, they have been subordinated to a strengthened central staff, and their access to the Minister of Defence has been reduced, while the power of the Chief of Defence Staff has increased.

In the decisions over Chevaline, the Navy's early preference was for Poseidon. The then Labour inner Cabinet overruled them, and later, when a Conservative government, which might have been more favourably inclined towards Poseidon, came to power under Heath, American reluctance to transfer the technology discouraged further exploration of the option.

It appears that the British military services generally do not make the running on the key nuclear decisions. They are, of course, consulted, and their views carry weight, especially on operational matters, submarine design, and similar specialised areas. The service chiefs and senior civil servants share the same assumptions on strategic nuclear issues and contribute to the consensus in favour of maintaining nuclear weapons.

*   *   *

In France, as in Britain, the military have played a subordinate role in the development of nuclear weapons. They were initially divided over whether France should acquire the nuclear bomb, fearing that it would divert money from conventional forces – as it did. Excluded from the informal networks based on membership of the Grand Corps, the military have had operational control of nuclear weapons, but not control of procurement. Even the prestigious 'Groupe S', which lays down long-term strategy and military policy, is chaired not by a military officer but by an engineer trained at the Ecole Polytechnique.

Within the armed forces, the nuclear commands have become a group apart. They are jealous of their privileges and compete with the conventional force commanders. Officers from these commands have won the greatest political influence, especially in appointments to the Private Staff of the President, the staff officers in the Elysée who are the President's closest source of advice and the bearers of the codes required to launch the French nuclear weapons.

In one instance, the military has lobbied strongly for the development of a new nuclear weapon. This has been the neutron bomb. The Army, in particular, sees this weapon as militarily useful against tanks, in contrast to the weapons of last resort operated by the Air Force and Navy. The Army also favoured the development of short-range, tactical missiles such as the 'Pluton' and 'Hades', although the political questions of how such weapons could be used without damaging West Germany have not been resolved.

*   *   *

Admiral Liu Huaqing is the Commander of the People's Liberation Navy (PLA), and a prominent figure in the modernisation of the Chinese armed forces. The Navy is developing submarines with nuclear missiles.

Liu was previously assistant to the Chief of Staff in the People's Liberation Army and a Rear-Admiral in the Navy. During the Cultural Revolution he was obliged to offer self-criticism and lost his post. He re-emerged in 1972 as Vice-Minister of Science and Technology and was given command of the Navy in 1982.

It is not easy to assess the role of the military in the Chinese nuclear weapons programme. After the Cultural Revolution, when the authority of the Communist Party was weakened by the attacks of the Red Guards, the PLA ran the government for some time. Later, when Lin Biao attempted a coup, the PLA was itself extensively purged. After the time of the Gang of Four, the PLA again became politically important, but recently Deng Xiaoping has cut back the size of the armed forces and reduced their political role. A number of military leaders have been removed from the Politburo. It remains to be seen whether these changes will last. Hu Yaobang, the General Secretary of the Party, said in an interview that, in order to get anything done in the military, he and the Prime Minister, Zhao Ziyang, had to use five sentences, while Deng Xiaoping achieved the result with one.

The PLA operates the Strategic Missile Force, and its Commander-in-Chief sits on the top-level Central Military Commission. The political commissars of the Strategic Missile Force are known to be recruited from the Public Security, China's secret police. There is no doubt at present that the Party exercises ultimate control.

*   *   *

'Thus far the chief purpose of our military establishment has been to win wars. From now on its chief purpose must be to avert them. It can have almost no other useful purpose.' These words were written by the American strategist, Bernard Brodie, shortly after Hiroshima and Nagasaki, when it became clear to everyone that the nature of war had suddenly and totally changed. However, the military establishments of the major nuclear states, which had reached positions of great political influence at the end of the Second World War, have consistently sought to fit nuclear weapons into traditional military goals. They accepted the idea of deterrence as a useful justification for their new position – 'Peace is our Profession', said the Strategic Air Command's logo. But the real concern of the military planners was to prepare for what would happen if deterrence broke down. In the United States and the Soviet Union, they planned to destroy the capacity of the other side to resist, by attacking each other's military forces. It is this military-based thinking which has led the superpowers to acquire such vast arsenals of accurate and lethal weapons. Military assumptions have come to permeate the process of nuclear weapons procurement, even when the military have not been directly in control of decision-making.

In the United States and the Soviet Union, the Military Services are in charge of the procurement of their own nuclear weapons and have an important role in development decisions. In the smaller nuclear countries, where for many years nuclear weapons were seen as a basically political weapon of last resort, the role of the military in nuclear matters has been less prominent than in the superpowers. In both France and Britain, the military do not control procurement and play a subordinate part in the choice of new weapons.

The military declare that nuclear weapons have maintained the peace and defended national security. If they are right, and they retain their present ascendancy, the world will continue to live in a tense and heavily-armed 'cold peace'. If they are wrong, the destruction visited on Berlin could be visited once more on a vastly greater scale.

5 The Defence Bureaucracies

[The role of the Ministry of Defence in the Chevaline and Trident decisions – the Pentagon, DSARC and the role of the civilian officials; the US Department of Energy and nuclear weapons – Soviet nuclear weapons procurement procedures and the role of the Defence Ministry – the National Defence, Science, Technology and Industry Commission of China – different roles of defence bureaucracies in East and West]

The Ministry of Defence is a long, white building just off Whitehall. It has been described as a 'desolate stone fortress' and 'the most impenetrable of all Whitehall bureaucracies'. Its chief distinguishing features are four battlement-like blocks containing offices, spaced at intervals along the top of the building, so that in places there are six storeys and in others nine. The entrance, off Horseguards Avenue, is dominated by two incongruous statues of bare breasted matrons.

On the sixth floor of the building are the offices of the heads of each of the Services – the First Sea Lord, the Chief of Air Staff and the Chief of the General Staff, as well as their overall boss, the Chief of Defence Staff. The Secretary of State for Defence also has his office here, next to that of his chief civil servant, the Permanent Undersecretary. So does the civil servant in charge of defence procurement, the Chief Executive of the Procurement Executive.

From this building the defence budget, running in 1985 at over £14 billion a year is managed and about half a million people are employed, two thirds of them in the armed forces and the rest civilians. From here too, the nuclear establishments at Aldermaston and the Royal Ordnance Factories at Burghfield and Cardiff are administered by an official called CERN (Controller of Establishments Research and Nuclear). Defence contracts for equipment, worth over £7 billion a year, are managed through the Procurement Executive. Co-ordinating them all are some 1000 senior civil servants who hold the top positions in the Ministry.

All the heads of the main defence bodies work in this building. However, their physical proximity does not make for a unified department. In practice the Services, the nuclear administrators and the top civil servants form separate communities. Each of the Armed

57

Services is in competition with the other two for its share of the overall
defence budget and much of the time of the senior Service officials and
civil servants is consumed in internecine fighting. Clive Ponting, a
former Assistant Secretary in the Ministry before his trial over the
Belgrano Papers, describes the Ministry of Defence thus: '. . . the
MoD really is divided against itself . . . you have constant battles
between the three services, particularly between the Royal Navy and
the Royal Air Force who really don't get on at all. You've obviously
got battles between the civilians and the military, particularly over
finance . . . Then of course you've got the scientists as well . . . all in all
it's a pretty much continuous conflict'.

Within the Ministry of Defence, the 'nuclear' officials are a distinct
and isolated group. The papers which they exchange have a higher and
more restrictive security classification than most which circulate within
the Ministry: 'Top Secret (Atomic)'. The nuclear side of weapons
procurement is handled by a committee called the Weapons Equipment Policy Committee (Nuclear). The papers from this subcommittee circulate to only 20 people.

The holders of the main nuclear posts in the MoD, such as CERN,
are normally recruited from the top posts at Aldermaston and are
thoroughly familiar with both the technical details of nuclear weapons-
making and with the personalities of the key scientists, technicians and
administrators. The importance of this group is clear in the history of
both the Chevaline and the Trident decisions.

In the case of Chevaline, the project was in the control of a narrow
circle of Scientific Advisers and 'nuclear' officials from its inception in
1967 to 1976. In this period, while the project was defined, weighed
against the alternatives, repeatedly reconsidered and finally approved
for completion, responsibility within the Ministry of Defence rested
with the Deputy Chief Scientific Adviser (Projects and Nuclear),
Victor Macklen. He reported to the Chief Scientific Adviser and after
1972 also to the Chief Executive of the Procurement Executive, then
Sir Michael Cary, on aspects which involved industrial collaboration.
Financial responsibility for the project was delegated by Sir Michael
Cary to Victor Macklen.

Victor Macklen has been described as Britain's 'Mr Nuclear'.
According to one report, 'as Deputy Chief Scientific Adviser at the
MoD he ran Britain's nuclear weapons programme, guarding his
knowledge, his empire and his unparalleled contacts in Washington
with a tenacity and secrecy which others in Whitehall found obsessive.
'In nuclear matters he had all the standing of a Marconi or a Baird',

says one insider, adding sourly, 'but he traded on it.' In action, Macklen was formidable. . . . 'The shock of white hair and the vivid rhetoric are indelibly etched on my mind', said one bruised American. Others in the MoD described Macklen as 'irascible' and a 'nuclear freak'.

In 1981, Sir Frank Cooper, the Permanent Undersecretary at the Ministry of Defence, was asked by the Public Accounts Committee about the responsibilities Macklen was given:

Q: Was it normal for the Deputy Chief Adviser (Projects and Nuclear) to undertake these sort of responsibilities?
Sir Frank Cooper: In a curious way this was an inheritance of the early days of nuclear work. It was kept for a very long period of time within an exceptionally limited circle of people who were privy to the innermost details of nuclear thinking and nuclear technology. As long as I can remember – which goes right back to the start – there was a somewhat similar arrangement to this where there was going to be a nuclear development.

Between 1972, when the 'project definition' stage was completed, and 1976, the Chevaline project proved more complex than anticipated. The Aldermaston scientists were having trouble designing the 'bus' at the front of the missile which would travel through space: this was an area strictly outside their competence. At the same time, the inner group of Cabinet Ministers who were considering the project were uncertain whether to commit themselves fully to the system and were financing it with short-term funds. Estimated total costs rose from £175 million to £594 million.

A review was then held within the Ministry of Defence. By this stage the Treasury was expressing doubts about the project. Nevertheless, the top men in the MoD decided it should go ahead. 'I cannot think of anything else that would have made anything like the same kind of sense', said Sir Frank Cooper. 'The Chevaline programme was looked at very carefully, alternative ways of meeting the same aim were looked at and Ministers came to the view, on advice, that we should go ahead with the Chevaline programme.' The advice came from Sir Frank himself, as well as the Chief of Naval Staff, the Chief of Defence Staff and the Controller of the Navy.

Control was taken out of the hands of the nuclear scientists, and given to the Navy. A Chief Weapons System Engineer under the Chief Polaris Executive took over the management and financial responsibility. British Aerospace was brought in as the contractor to manage the project. The civil servants privately thought that the nuclear scientists had made a mess of the project, though in his evidence to the Public Accounts Committee, Sir Frank Cooper argued that they were 'doing their best in the light of their knowledge at the time'.

A further, final review was carried out in 1977. Once again the advice of the top civil servants and military was to continue the project to completion. Ministers took the advice.

Sir Frank Cooper was Permanent Undersecretary from 1976 to 1982. Educated at Manchester Grammar School and Oxford, Cooper was an RAF pilot in the war. He then joined the Air Ministry and spent most of his subsequent career in the Air Ministry and then the Ministry of Defence. He has been described as 'the most powerful constant factor in the defence establishment'. Having left office, he expressed the view that Britain would have been better off if it had been decided not to have nuclear weapons originally, but having decided to have them, it was better to keep them. In office, however, he staunchly supported the deterrent. 'You know, people go on and on about nuclear war', he said, 'millions and millions of people died in the 1914 war, it just took longer to kill them, that's all.'

He delegated most of the nuclear decision-making to others but his role was important, as a channel of advice to Ministers and as a maker of appointments. In 1977, in readiness for the forthcoming decision on Trident, before the Labour Government had taken any decision, Cooper appointed a new Deputy Under Secretary (Policy), Michael Quinlan. Quinlan was to become the key civil servant while the Trident decision was undertaken.

Quinlan was educated at Wimbledon College and Merton College Oxford, where he took a First in Mods. A convinced Christian and very bright, he was fascinated by what the civil servants call 'nuclear theology' – the arcane logic of deterrence strategy. He liked the intellectual challenges it involved.

The decision-making process on Trident began in 1977 and ran on at least until 1982. Even before 1979, when the new Conservative government headed by Margaret Thatcher came to power, the decision began to be prepared within the Ministry of Defence. In 1977 the Navy warned that the hulls of the Polaris submarines might need to be replaced by 1990. A decision about a successor system would therefore be required in 1980.

There was never any doubt or debate within the Ministry of Defence about whether Polaris would be replaced. The only question was what to replace it with. David Owen, the Foreign Secretary, favoured a submarine-launched cruise missile system, and invited American experts to London to discuss it in 1977. Victor Macklen, who favoured Trident, attended the briefing. He was scathing in his criticism: 'we do have some experience shooting down cruise missiles you know. We used to call them V1s'. Macklen had begun his career doing research into countermeasures to the V1 buzz-bombs in 1944 and 1945.

Two teams were set up to examine the question, one under Sir Anthony Duff in the Foreign Office and one under Sir Ronald Mason, the Chief Scientific Adviser in the MoD. Their studies were hypothetical and secret, since the majority of the Labour Party which was then in power were opposed to replacing Polaris. The Prime Minister, James Callaghan, personally authorised these studies, so that the Secretary of State for Defence, Fred Mulley, who was a member of Labour's National Executive, could say he had authorised no such thing. Owen and the Foreign Office managed to get Victor Macklen excluded from the study team, but nevertheless his influence on the Mason report was substantial.

The Mason/Duff report carefully avoided recommending any specific system, but its analysis pointed clearly towards Trident. It was considered by an inner Cabinet committee, consisting of the Prime Minister, the Foreign Secretary, the Secretary of State for Defence and the Chancellor of the Exchequer, in secrecy from the rest of the Cabinet.

This inner group was divided and no decision was taken. Nevertheless the Prime Minister, Callaghan, raised the matter of the availability of Trident with President Carter in December 1978, as they strolled along the beach on the Caribbean island of Guadeloupe. Carter said he had no objections to offering the system to Britain.

In May 1979, Labour lost the General Election and the inner group of Labour politicians was relieved not to have to take the decision.

With the Thatcher government in power, the outcome was not in doubt. The decision was taken quickly, on a very informal basis. A strong consensus in favour of Trident had already emerged in the MoD. Quinlan wrote down the arguments in favour in his crabbed hand-writing. Most of the work was done in a committee of civil servants attached to the Cabinet Secretariat, in which Quinlan took the lead. The Mason/Duff papers were used as the basis. The papers went to the MISC 7 Cabinet Subcommittee, which included Mrs Thatcher (Prime Minister), Mr Howe (Chancellor), Mr Pym (Defence), Lord Carrington (Foreign Secretary) and Mr Whitelaw (Home

Secretary). They endorsed the decision to go ahead in general terms. The civil servants proceeded to work out the details.

Most of this was done by Quinlan. He handled the negotiations with the US government with the help of the Foreign Office. A series of top-secret negotiating sessions were held, one in an immigration office at Heathrow Airport and another at an exclusive restaurant on the Champs Elysées in Paris. Part of the agreement was signed by Quinlan and his American counterpart on the boot of a car in a car-port of the British embassy in Washington.

Formal letters reiterating what Quinlan, the Foreign Office and the Americans had worked out were then exchanged between President Carter and Mrs Thatcher.

In 1982 a further major decision had to be taken, when the US Government decided to phase out Trident I missiles in favour of the more powerful Trident II. Although this meant a large increase in cost and the acquisition of more destructive power than had been considered necessary, the Ministry of Defence and the government had little hesitation in proceeding to follow the Americans and ordering Trident II.

The decision having been taken in general terms, it remained to work out the technical details. The Chief of Strategic Systems Executive was given this responsibility, reporting to the Controller of the Navy. An operational requirement was written and the request for the submarine went through the formal process of the Operational Requirements Committee and the Defence Equipment Policy Committee. The Director of Naval Warfare gave his views on operational matters such as what level of quietness and range would be desired. Knowing the Trident programme had top priority, the Navy proceeded to ask for the latest and best sonar and a new pressurised water reactor to be developed for it, which they wanted for their other submarines. With an existing commitment to the deterrent being 'the best available', these requests were agreed. Not to be left behind, the nuclear scientists won permission to substantially rebuild and modernise the centre of Aldermaston with a complex of facilities intended not only for the manufacture of the Trident warheads but also for further research into subsequent generations of nuclear weapons.

In this way the crucial decisions were taken, largely behind closed doors within the Ministry of Defence, in the Cabinet subcommittee MISC7 and in the Cabinet Secretariat. The papers were secret even to most of the top civil servants. Only a narrow circle considered them. Abandoning the nuclear deterrent was not presented or even considered as an option. Given the assumptions this inner circle shared, the decision seemed inevitable.

The decision-making style reflects the secretive and closed nature of the MoD. 'It is a very claustrophobic world', says Clive Ponting. 'The MoD is very different from the rest of Whitehall. It tends to have most of its contacts with the Foreign Office who tend to be dealing with the same sort of material. MoD doesn't have much contact with the outside world. It's talking very much with the American Department of Defense or NATO . . . it is a very claustrophobic atmosphere in there.'

Civil servants are recruited from similar backgrounds and share similar assumptions. Eighty-five per cent are educated at Oxford and Cambridge. People of acceptable views are selected first at the initial interview and later through the promotion process. All the civil servants in the 'fast stream', who are promoted to the top jobs in the Ministry, undergo 'positive vetting' in which their political beliefs and past associations are investigated. In order to reach the highest positions, it is essential to conform to the shared beliefs and culture of the Ministry. Ponting says, 'I think there are certain shared assumptions inside MoD that are not discussed and often not really written down.' These are that the nuclear deterrent is essential, that the special relationship with the United States of America is all-important, that there is a Soviet threat, that Britain has a role outside NATO, and that Britain stands in a special position between the US and Europe, and when the chips are down must support the US. There is no internal debate on these matters, and the MoD's policy-making is unaffected by the public debate outside.

The top civil servants spend most of their working lives in the Ministry. Although they rotate around within the Ministry, they rarely move outside it. 'They tend to become fairly conservative with a small 'c' people, interested in the status quo, often very friendly with the military and they tend to absorb their sense of values.'

At the top of the Ministry are the civil servants, whose skill is in paperwork, administration, and policy. They appoint others of the same mind to take their places, and tacitly confine the role of the military, the scientists and the economists to giving specialist advice. Politicians, too, are kept out until it is necessary or desirable to consult with them. Sir Frank Cooper was asked, 'How do you think that defence management should be approached ideally?' He replied, 'Well I think that the less the Government does inside the Government the better, because you cannot have efficient public sector management where ministers are accountable for everything their so-called managers do.'

All this makes the Ministry of Defence very impervious to change. It can only be controlled from the top, yet it is difficult for a Secretary of State to establish effective control. Ponting was asked, 'Because of its size and its range of responsibilities, the MoD appears to be very difficult to control. Is anyone really in charge?' He replied, 'Not really, no. I think the answer has to be no. In many ways the machine, and it is a machine, is running on auto-pilot. It's set going in one direction and all the pressures keep it going in that direction. The chances of one person actually influencing it are fairly small. They're not usually in one place for very long and on something like a complex weapons system the time-scales for decisions are enormously long, 10, 15 years, so that even one minister is probably only going to see a project once in his time inside the Ministry of Defence. His chance to influence it is pretty small. So what you have really is a military machine and an equipment industry machine all going in one direction.'

*   *   *

In no country has a bureaucracy exercised greater power over nuclear weapons decisions than in France. In the early stages of the French nuclear weapons programme, during the unstable period of the Fourth Republic, successive politicians were either opposed to the French nuclear weapons programme or ambivalent about it. All the drive came from the elite group of engineers and technocrats in the CEA. From his office as director, Guillaumat pushed the nuclear weapons programme forward in a period which saw 11 Prime Ministers come and go. When asked which of the politicians who were opposed to the weapons programme did most to obstruct him, Guillaumat replied: 'No one ever lasted long enough.'

At the time General de Gaulle came to power, bringing a strong Presidential commitment to the 'force de frappe', the nuclear weapons programme was already strongly established in the government.

Besides the Directorate of Military Applications in the CEA, the other leading organisation is the Directorate General for Armaments, part of the Ministry of Defence. Like the CEA, it is dominated by engineers and graduates of the Grandes Ecoles. It is not surprising that this group too shares the same values. They believe that decisions on nuclear matters are best taken by a well-qualified élite within the top levels of government. They are committed to an advanced level of military technology, in the interests of a strong French state, and they unquestioningly support the development of nuclear weapons.

*   *   *

Although the British and French Ministries of Defence are large bureaucracies, they are dwarfed by the Pentagon. Situated on the bank of the Potomac River, this fort-like building has become a symbol of the US military establishment. It consists of not one but five concentric pentagonal buildings, each of five storeys. It is built of granite, in a neo-Prussian style, with rows of featureless windows along its 920-ft outer sides. A total of 23 000 people work in its offices and its corridors run 17.5 miles. Nearby there is a helicopter pad and a 67-acre parking lot.

The Pentagon was built by the same General Groves, of the Army Engineers Corps, who later took control of the Manhattan Project to build the first atomic bomb. It was intended that after the war it should become a hospital. Instead it has remained in constant military use, becoming the headquarters of the Department of Defense when the Department was created by the National Security Act of 1947. The Pentagon houses both the civilian officials of the Department of Defense and the Chiefs of Staff and leading officials of all the Armed Services.

The inside 'E' ring, which looks on to the tree-shaded courtyard in the centre of the Pentagon, is reserved for the Secretary of Defense and his most senior staff.

In the basement of the 'E' ring, every two weeks, six men meet in a conference room and decide to spend hundreds of millions of dollars on new weapons systems. This group is known as the Defense Systems Acquisition Review Council (DSARC). Its members include the Undersecretary of Defense for Research and Engineering, the Undersecretary of Defense for Policy, the Comptroller, the Director of Program Analysis and Evaluation, the Chairman of the Joint Chiefs of Staff and the Assistant Secretary of Defense for Manpower and Logistics.

This council reviews research and development programmes costing more than $200 million and all production contracts costing more than a billion dollars. These major new weapon projects have to pass through several formal stages. The first is the 'DSARC Milestone

One', which establishes that there is a military need for a weapon and approves the start of development work. Until recently 'DSARC Milestone Two' had to be passed before a weapon could enter full-scale development, but now the Services are permitted to proceed with full-scale development before obtaining DSARC approval. However, DSARC Milestone Two must be passed before proceeding far. On paper, there is a DSARC Milestone Three which authorises a weapon passing into production. The Reagan administration has dispensed with this Milestone to enable the Services to speed up their procurement.

Running in tandem with the acquisition process is the budgetary cycle, whereby funding for weapons projects is approved. The full cycle for each financial year takes 21 months. It starts with the Secretary of Defense outlining the threats to the nation's security in a document called the 'Defense Guidance', which is prepared on the basis of assessments by the Joint Chiefs of Staff. The services then recommend the weapons and forces they wish to be funded in Service budget plans. These are reviewed by the Office of the Secretary of Defense and modified as necessary to produce the Secretary of Defense's budget request. This is then reviewed by the Office of Management and Budget and by the White House, and is put into the President's budget request. Finally, the Congress sets a ceiling for the total defence budget, and then drafts a law authorising individual programmes.

In order to assist with the planning of the budget, the civilian managers of the Defense Department use a computer programme called the Planning, Programming and Budgeting System. This is intended to relate military plans to budgets and to enable constant reformulation of the budget as weapons programmes are adjusted throughout the year.

The civilian officials in the Department of Defense have the power to control major weapons projects through both budgeting and acquisition decisions. The extent to which they exercise this power has varied in different administrations, but an important feature of their position is that in both the budgetary and the acquisition cycles their role is to react to Service initiatives.

Constitutionally, the civilian officials in the Pentagon have authority over the military. This authority is exercised from the Office of the Secretary of Defense. Beneath the Defense Secretary is a Deputy Secretary, two Undersecretaries (one for Policy, one for Research and Engineering) and eight Assistant Secretaries. All of these top jobs are political appointments, unlike in the British Civil Service, and the incumbents rotate with changes of Administration.

This means, in principle, that the top officials reflect the will of the Administration in power. It has also meant a rapid succession of different officials in the top posts. Since 1945, the average period in office has been, for the Secretary of Defense, 2.3 years, for the Deputy Secretary, 1.8 years, for the Undersecretary for Policy, 1.7 years, for the Undersecretary for Research and Engineering, 4 years. Because of this 'personnel turbulence', the civilian officials have rarely exercised effective, continuous control over major weapon systems which take ten or more years to develop. It is possible for an energetic civilian official to slow down and block a weapons programme, but extremely difficult to cancel it outright, as the military can come back with a request for a weapon year after year.

Another consequence of the rapid succession in top jobs is that the civilian officials often have to train 'on the job'. Some come to defence posts from other fields; given the short average time in office they tend to rely heavily on the professional advice of the military. Others come from the weapons laboratories or from jobs with the defence contractors. They tend to be well informed but already heavily committed to the defence community's assumptions and way of thinking.

The key official in the Department of Defense with responsibilities for developing nuclear weapon systems, although not their nuclear warheads, is the Undersecretary for Research and Engineering. It is in his office that the different requests of the Services, intelligence information, data from missile tests and arms control considerations are brought together, and it is from here that a course is plotted through the buffeting cross-pressures of the Services, the Congress and the White House. Almost all of the holders of this position have been recruited from either the nuclear weapons laboratories or the aerospace industries.

From 1965 to 1973 the holder of this post (then called Director of Defense Research and Engineering) was John Foster. Foster was instrumental in promoting the MIRV programme and suggested to the Navy that they should look into strategic cruise missiles in 1971. 'Foster took it for granted that technology should be pushed as hard as possible, although he recognised the need to choose from the wide variety of different possible new technologies. He also believed that when technology reached the stage where it was militarily effective, it should be deployed.'

Beneath the Undersecretary of Defense Research and Engineering are a number of other important posts, including that of the Deputy Undersecretary Research and Engineering (Strategic and Theater Nuclear Forces). In 1985 this was held by Thomas K. 'T.K.' Jones. With florid grey hair and watery eyes, 'T.K.' has supervised the procurement of a new generation of accurate counterforce weapons. Born in 1933, he worked for Boeing from 1954 to 1971 on ICBMs, AMB systems, strategic bombers and manned space vehicles. From 1971 to 1974 he served on the US delegation to the SALT talks. He then returned to Boeing as Deputy Requirements and Strategy Planning Manager before taking up his post in the Undersecretary of Defense Research and Engineering Office under the Reagan Administration. Obsessed by the Russian threat, Jones is particularly keen that the US should emulate the Soviet civil defence system. He is responsible for the widely-quoted remark that if only the American people had enough shovels to dig themselves holes to shelter in, the United States could survive and win a nuclear war and be on the road to recovery in two years.

The Department of Defense is not the only government agency involved. An important section of the bureaucracy which develops nuclear weapons operates from the Department of Energy. It is this department, not the Department of Defense, which is responsible for administering the nuclear weapons laboratories and the manufacture of nuclear warheads, as well as the production of the nuclear materials required for the weapons programme. In line with a statutory obligation under the 1946 Atomic Energy Act, the Department of Energy operates its own system for classifying nuclear information, the 'Q'-clearance system. This is a separate and more restricted classification than Department of Defense clearances.

The agency inside the Department of Energy responsible for nuclear weapons is the Office of Military Applications. This office liaises with the Department of Defense through a Military Liaison Committee, which advises the Department of Energy on military requirements. The Director for Military Applications in the Department of Energy and the Chairman of the Military Liaison Committee are two of the most powerful people in charge of US nuclear weapons.

Major General William Hoover was the Director for Military Applications in the Department of Energy in 1985. He is a fair-haired, square-jawed clean-cut Air Force General. Born in 1932, he graduated from the US Naval Academy and then became a pilot in the Air Force. During the Vietnam War he was a Wing Commander at Da Nang and flew 97 combat missions, winning several decorations for distinguished service. He served as a technical manager for various research and development projects at the Air Force Space Systems Division in Los Angeles and later was a planning and programming officer for the Deputy Chief of Staff of the Air Force. He also had a spell at SHAPE – the Supreme Headquarters Allied Powers Europe. He now directs the US nuclear weapons programme and has management responsibilities for the research, development, testing and production of nuclear weapons.

Dr Richard Wagner was appointed Chairman of the Military Liaison Committee in 1981 and Assistant Secretary of Defense (Atomic Energy). Born in 1936 in Oklahoma City, Wagner took a Ph.D. in physics and worked on nuclear weapons at the Livermore Laboratory from 1963 until 1981. He designed a warhead for an anti-ballistic missile and was a Division Leader and later Associate Director at Livermore. For several years he was in charge of nuclear tests at the Nevada test site. He has served as an adviser to the Joint Strategic Target Planning Staff and sits on the Defense Science Board which advises the Secretary of Defense on weapons programmes. Wagner has fair hair, deep-set eyes and a goatee beard. He is married with two sons and a daughter.

The Defense Science Board is one of a number of top-level advisory boards and study panels which are used extensively to prepare reports on specific systems. Its members are scientists from the weapons laboratories, industrial managers from the defence contractors, retired military men and a smattering of other industrialists and academics. This committee advises the Undersecretary for Defense Research and Engineering on new weapons projects. There are similar Advisory Boards for the Armed Services.

Ad hoc panels of similar composition are called from time to time to advise on new weapons, especially when programmes have reached moments of crisis. One such panel was the Scowcroft Commission, formed to discuss the future of the troubled MX missile. Another was the Joint Strategic Bombing Study, which considered the future of the B-1 bomber.

Sandwiched between advisory bodies and the military Services, the civilian officials within the Department of Defense have less power than their French and British counterparts. The civilian officials can speed up or slow down the Services programmes and occasionally – as in the case of Cruise – promote a new programme which the Services do not want. The political appointees can, briefly, stamp the character of the Administration on their departments. But it is not easy for them to block programmes the military want. Hitherto they have rarely been a force for restraint.

*   *   *

Unlike its western counterparts, the Soviet Ministry of Defence is entirely military in composition. It contains no civilian officials, and its voice is the voice of the Soviet Armed Services within the government. One of the First Deputy Ministers is also the Chief of General Staff; another is the Commander-in-Chief of the Warsaw Pact Forces. All the civilian control comes from above – from the Defence Council and the Politburo.

There is a formal procedure for the development of all new weapons, with clearly-defined stages, which is applied to all types of military technology. The first stage is Project Identification. A new project can be initiated by a design bureau, a research institute, one of the Technical Administrations or Scientific–Technical Committees of the Defence Ministries or of the Armed Services, or by the General Staff. The design bureau, in consultation with the military customer (such as the Strategic Rocket Force) then prepares an 'Advanced Project Specification', which is evaluated by the Scientific–Technical Council of the design bureau's Ministry and by the military customer. At this point a decision is made as to whether to proceed to prototype design. The Military Industrial Commission, the Defence Council and the Politburo – all bodies senior in status to the Ministry of Defence – are involved in this decision.

If the project is approved, a 'Tactical–Technical Assignment' is drawn up by the Technical Administration of the Service customer. This sets out the military and economic requirements for the design and includes an estimate of the volume of production. This document is used to monitor the progress of the project under the design bureau. The design bureau then prepares a preliminary Draft Design. This is evaluated by the Scientific–Technical Council of the Design Bureau and by research institutes of the Defence Industry Ministry concerned and by the military customer. If approved, the Design Bureau prepares working drawings and builds prototypes of the weapon. These are then tested, first in the factory and then by a State Commission headed by the military customer, to check that they meet the requirements of the Tactical–Technical Assignment.

Then there is a second key decision, on whether to transfer the prototype to series production. Here again the Military–Industrial Commission, the Defence Council and the Politburo decide whether the system is to proceed. The Defence Industry Ministry – for missiles it would be the Ministry of General Machine-Building – then proceeds to manufacture the weapon, with assistance from personnel from the design bureau. Further testing is carried out by the military and the decision to deploy is then taken by the Armed Services, the General Staff and the Ministry of Defence in conjunction with the Defence Council and the Politburo.

In this process, the military, through the Ministry of Defence, place orders to procure new weapons and supervise their development, testing them at key points before they enter into service. The Ministry of Defence thus has a crucial administrative role, and it monitors each stage of the weapon's cycle in detail. But overall control at the decisive points of entry into prototype design and entry into series production are taken by the central policy-making bodies. The Ministry of Defence has an important role in nuclear weapons decision-making but it appears to be less central than the US Department of Defense, the British MoD or the French Defence Ministry.

* * *

In China, the Ministry of National Defence has less power over nuclear weapons than in all the other four major nuclear weapon states. It is probably one of the world's least important defence ministries, especially considering the size of the military forces China commands. For many years the Ministry was dominated by nominally lower institutions such as the General Staff. The Minister is important, in his capacity as First Vice-Chairman of the Central Military Commission – a top central body. But the role of the Ministry is purely administrative, and its importance in nuclear matters appears to have been negligible.

A more important government department in nuclear matters is the National Defence, Science, Technology and Industry Commission. This body, formerly subordinate to the Party Military Affairs Commission but now under the State Council, has had overall direction of the nuclear programme. It controls the allocation of defence research and development funds and co-ordinates the development of nuclear weapons. Its Chairman, Fang Yi, is a member of the Politburo and of the State Council. Its vice-chairman in 1981 was Qian Xuesen, the rocket engineer. Nie Rongzhen, who directed the nuclear weapons programme in the early years, was a predecessor of Fang Yi's as chairman of the National Defence, Science, Technology and Industry Commission. During the Cultural Revolution he was accused of having founded an 'independent kingdom' in this Commission, but, with Zhou En Lai's protection, Nie survived and the nuclear programme was insulated from the political turmoil.

*   *   *

There is a striking contrast in the relative importance of the permanent defence bureaucracies in the five major states. In the centrally-planned countries, perhaps because the senior political bodies are themselves long-standing and bureaucratic in nature, the defence ministries have less autonomy over nuclear decision-making than in the western countries. In the USA, France and Britain, the defence ministries have great power. The Pentagon and the military section of the Department of Energy dominate nuclear decision-making within the government, but within the Pentagon the Services hold a dominant position. In France and the UK, it is the civilian officials with long terms of service, operating within closed and secretive bureaucracies, who have come to take a substantial degree of control of the development of nuclear weapons.

Power has accrued to institutions in which the longevity of terms of service matches the timespan of the weapon's development period. In the West, power over nuclear weapons has tended to lie with the permanent bureaucracies, either civilian or military, who do not trust the politicians. In the East, where the Politburos have great continuity, power lies with the politicians, who do not trust the bureaucrats.

6 Presidents, Prime Ministers and Politburos

[The British Inner Cabinet – the US President and inter-agency committees – the Soviet Politburo – the Chinese Central Military Commission – the French President and the Conseil de Defense – differences in political control]

Three metres below ground level, underneath the Government Offices in Great George Street, off Whitehall, are the Cabinet War Rooms. In these blast-protected headquarters, the Government directed British operations during the Second World War at the times when London was under air attack. The visitor can tour the premises today and get an idea of the nature of the centre of government as it was in war-time during the 1940s. The War Cabinet met in the Cabinet Room, around tables arranged in a square. Churchill sat in a large wooden chair in front of a map of the world. Eight ministers attended these meetings, together with top military officers and officials of the Cabinet Secretariat. The same room was used for the Defence Committee – an inner group comprising only Churchill, Attlee (the Deputy Prime Minister), the Cabinet Secretary and the Chiefs of Staff. Nearby there is a room containing a display of the war situation on large wall maps, with an array of coloured telephones, some with scramblers fitted, which connected the Cabinet to government departments, the Admiralty and the War Office. Another room contains the personal radio-telephone link Churchill used to talk to the President of the United States, routed via a large, electronic scrambling device located in the basement of Selfridges. From the broadcasting room, Churchill transmitted messages to the nation. A typing pool with Remington manual typewriters kept up with the flow of paperwork, and the staff were accommodated in a number of offices, messes and bedrooms.

Churchill relied on a very small group of ministers to make military decisions and he took over the Defence Ministry himself. He took decisions about Britain's war-time collaboration with the USA in the making of the atomic bomb with a very small inner circle of advisers during this period. This style of decision-taking, which developed during war-time exigencies, still prevails today in nuclear matters, remarkably unchanged.

In Britain, the most important decisions on nuclear weapons are taken by an inner circle of senior politicians and senior civil servants. These matters are not discussed by the whole Cabinet, but rather by an inner subcommittee of the Cabinet. The meetings of this committee are not publicised and details of the discussions are not available. MISC 7 was an *ad hoc* committee set up to consider replacements for the Polaris submarine. A similar subcommittee was used for discussing Chevaline.

Normally, the members of these Cabinet subcommittees are the Prime Minister, the Secretary of State for Defence, the Foreign Secretary and the Chancellor of the Exchequer. The meetings are irregular. They are called by the Prime Minister on the recommendation of the Cabinet Secretariat. Briefing papers are prepared by the civil servants and these lay out the range of options within which Ministers take their decisions.

Some idea of the character of such briefing papers can be gleaned from those used for the initial decision to build a British bomb during the Attlee Government of 1945–50. The options were laid out for Attlee's inner Cabinet by Lord Portal (Chief of the Air Staff during the war, a veteran of Britain's Strategic Bombing Offensive and then Controller of Production, Atomic Energy, in the Ministry of Supply – a post especially created by Churchill):

'I submit that a decision is required about the development of atomic weapons in this country', began Portal. Margaret Gowing, the official historian of Britain's atomic programme, summarises the paper as follows:

He suggested that there were three courses: (a) not to develop atomic weapons at all; (b) to develop the weapon by means of ordinary agencies in the Ministry of Supply and Service departments; and (c) to develop the weapon under special arrangements conducive to the utmost secrecy. He imagined (a) would not be favoured by the Government in the absence of international agreement, while if (b) were adopted it would be impossible to conceal for long the fact that the development was taking place. Moreover it would certainly not be long before the United States authorities heard that Britain was developing the weapon, and this might seem to them another reason for reticence over technical matters, not only in the field of military uses of atomic energy but also in the general knowledge of production of fissile material. If, for national or international reasons, the special arrangements at (c) were required, they could well be made through (the secret organization set up under Lord Portal). . . . Portal therefore asked for directions on two points – whether research and development on atomic weapons were to be undertaken and, if so, whether the special arrangements were to be adopted.

A single meeting of the Cabinet subcommittee GEN 163 considered Portal's report in conditions of great secrecy in January 1947. The meeting decided that work on atomic weapons should be undertaken and approved the special arrangements set up by Lord Portal.

Some time later, the only paper suggesting that Britain should not develop atomic weapons, written by Professor Blackett, was circulated to a Cabinet subcommittee. Apparently Blackett had not been aware that the decision had already been taken.

This style of decision-making, using secret *ad hoc* committees of Cabinet, with briefings from senior civil servants, continues to the present day. The character of the briefings over Chevaline and Trident will not be made public for 30 years. Presumably the civil servants' briefs layed out certain options and made a recommendation. Brian Sedgemore MP writes of the Cabinet Secretariat's briefings: 'The Chairman's briefs (that the civil servants) draw up are powerful documents often leading in a certain and clear direction – often in the direction thought best by civil servants who have (held parallel meetings to service the committees).'

The decisions on Chevaline were kept a secret from the whole Cabinet, which never knew in any detail what was going on. The Cabinet was told in general terms that the Polaris missiles were to be improved, but even the code word 'Chevaline' was unknown to Ministers outside the Cabinet subcommittee involved. In her diaries, Barbara Castle describes the meeting of the full Cabinet in 1974 at which the programme was mentioned:

The main rub came over nuclear policy, on which Harold [Wilson] was clearly expecting trouble. He needn't have worried: Mike's [Foot's] comments were so muted as to be almost token. Harold prepared the way carefully by saying that, although we would keep Polaris and carry out certain improvements at a cost of £24 million, there would be no 'Poseidonisation' and no MIRV. The nuclear element represented less than 2 per cent of the defence budget but it gave us a unique entrée into US thinking and it was important for our diplomatic influence for us to remain a nuclear power . . . Mike came in almost hesitantly. He admitted Harold was trying to keep within the compromise of the manifesto on this, though we were committed to getting rid of the nuclear bases. 'We shall proceed to negotiate this within the overall disarmament talks', Harold countered promptly. Mike then said that he remained of the view that we should rid ourselves of nuclear weapons, but recognised that he was in a minority and so would not press the matter. Peter [Shore] and Wedgie [Benn] said nothing. I was more emphatic than Mike. . . . The debate then died away. Harold summed up cheerfully, saying that Cabinet, with a few of us expressing dissent, had endorsed the policy.

On Trident, too, only a very narrow circle of Ministers was included in the subcommittee, MISC 7, which took the decision. Besides the Prime Minister, the Secretary of Defence, the Foreign Secretary and the Chancellor of the Exchequer, only William Whitelaw, Mrs Thatcher's deputy, was included. The briefing papers were prepared by the Cabinet Secretary, Sir Robert Armstrong, and by a team of civil servants including Michael Quinlan from the Ministry of Defence, who took the lead. The option of not replacing Polaris was not even considered.

Thus, British nuclear decisions have been in the hands of a tiny, inner circle of top Ministers and civil servants. These people have considered the decisions too secret and too important to take even the whole Cabinet into their confidence, let alone a wider circle of opinion. Hitherto, until 1985, there has been a striking continuity of nuclear policy among the changing members of this inner circle, despite the policy differences expressed in the party manifestos. The top politicians of both parties have preferred to keep their nuclear deliberations secret from the public. When, in an unusual departure from this policy, Mrs Thatcher's Secretary of Defence, Francis Pym, revealed the fact that £1000 million had been spent on Chevaline without Parliament being told, the former Labour Defence Minister Fred Mulley said, 'it is one of the most outrageous, disgusting, most damaging examples of breaking the continuity of nuclear decision-making there has ever been'. His angry remark offers a striking insight into the attitudes of the inner circle which, in general, has observed its own unwritten rules.

One other source of influence on British nuclear decisions is of decisive importance. This is the United States. Both directly and through NATO, the influence of the United States on British nuclear policy has been crucial. It was the United States which in 1962 cancelled the Skybolt missile, which had been promised to Britain, and offered Polaris instead. The Nixon government's refusal to offer Britain Poseidon led Heath's government to proceed with Chevaline, and later the Carter government's willingness to offer Britain Trident helped to make the decision a *fait accompli.* Later when the American government decided to phase out Trident I in favour of the more expensive and longer-range Trident II, the British government felt it had no option but to follow suit and change its order to Trident II.

Two aspects of the 'Special Relationship' are crucial for Britain's nuclear role. One is the 1958 Mutual Defence Agreement which provided for the sharing of nuclear design information and nuclear materials. The other is US intelligence, which provides essential operational information for the British nuclear force. In return British Governments have been willing to provide military bases on UK soil for US nuclear forces and consistently to support US positions on arms control and arms deployments. Sir Frank Cooper, the Permanent Under Secretary of Defence, said, 'if you ask me whether the Americans have an undue degree of influence over British defence policy I would have to say yes.' Clive Ponting was asked in an interview, 'Are you saying that we are in fact a client state of America?' Ponting replied, 'Client state is putting it a bit strongly but there are very clear signs I think that it's not far short of that. . . . They clearly do have an undue degree of influence because when the chips are down we side with the Americans because we think the American nuclear and intelligence material is so important to us that we are prepared to pay that price to keep that material flowing.'

In the last analysis, British decision-making on the independent nuclear deterrent turns on US decisions.

*   *   *

The 39th President of the United States of America, Jimmy Carter, was the first to have personal experience of nuclear matters. Carter had trained as a nuclear engineer and had held command in a nuclear submarine. He understood the details of nuclear weapons better than most Presidents and took a keen interest in the plans for their development and use.

Carter came to office intent on doing something to restrain the 'mad momentum' of the arms race. During his campaign he had spoken out against the Air Force's B-1 bomber and had called for a cut-back in military expenditure. He was suspicious of the Pentagon's unending demand for new weapons and determined to seek some restraint by negotiating a second Strategic Arms Limitation Treaty with the Soviet Union.

Three years later, Carter discovered some of the limitations of his power. He had cancelled the B-1 bomber, but in its place he had authorised the construction of a huge number of air-launched cruise missiles to be carried by the B-52 bombers. Although he had once pointed out that a single nuclear missile submarine would be sufficient to deter the Soviet Union from launching an attack on the United States, and during the presidential campaign he denied the possibility of limiting a nuclear war to selected strikes, in office he issued Presidential Directive 59, which directed the Pentagon to ensure that the United States would be capable of fighting a protracted nuclear war. He had called the Air Force's ideas for deploying the new MX missile in a 'shell game' by moving them from one silo to another 'the craziest thing I ever heard', yet he ended up authorising the development of the MX missile. He also approved the new Trident submarine with its highly accurate MIRVed missiles. In this way, Carter signed his name to most of the programmes that have become associated with the Reagan military build-up. His one dramatic measure of restraint, the cancellation of the B-1, was later reversed. Directly after the cancellation, the Systems Command of the Air Force continued to work on R & D programmes on four pre-production aircraft which had been delivered to Edwards Air Force Base. When the Reagan Administration came to power, the Strategic Air Command lobbied successfully for the reversal of Carter's decision.

Carter had found that it was neither possible to defeat the Pentagon nor politically desirable to take them on. With his foreign policy badly damaged by the hostage crisis in Iran, he needed a SALT agreement with the Soviets in order to sustain his prospects for re-election. But there was no hope of persuading Congress to sign a SALT II treaty if the Joint Chiefs of Staff testified against it, and Carter could not get the support of the Joint Chiefs of Staff unless he agreed to their cherished weapons programmes. In meetings at Camp David and at the White House, Carter was persuaded of the military necessity for a modernisation of the Minuteman missiles, which intelligence predicted would soon become vulnerable to a Soviet attack, and of the bomber and submarine legs of the Triad. The 'doves' in his Administration, anxious for the Administration's backing for the SALT treaty, decided not to resist the new weapons programmes. In any case, as the Administration went on, they were forced out of their positions by more 'hawkish' counsels. Carter's own mind changed as he spent more time in office, briefed by the Pentagon and the National Security Council in place of his own more liberal advisers. He began to accept the necessity of the military build-up. As his term wore on, he looked older, greyer, and more resigned.

The President has the constitutional authority to intervene in nuclear weapons decision-making at almost any point. President Reagan's personal commitment to the Strategic Defense Initiative has shown that a President has the power to launch an entirely new weapons initiative, even if, as the details are worked out by the weapons laboratories, the contractors and the Pentagon, the programmes which proceed differ in character from the rhetoric which launched them. Normally, however, the President's involvement with nuclear weapons programmes is in the context of broader considerations such as the budget process and the framing of arms control agreements. Through his powers of appointment and his speeches, the President can set the style and direction of an Administration. The President also formally endorses the acquisition of every nuclear warhead, by personally signing the Annual Stockpile Memorandum. Otherwise, the President has no regular or routine involvement with decisions to develop nuclear weapons. In practice, the Presidency is involved in major decisions over the course of particular programmes only when they have reached a point of crisis, or when other protagonists in the decision-making process are divided and choose to submit the decision to Presidential arbitration.

A troubled programme such as the MX missile, for example, was submitted to the President for a decision on several occasions. When the Carter Administration came into office, it cut the funding of the MX programme for two years in succession before agreeing to full-scale development in 1979. In June 1979 a formal meeting of the National Security Council was held in the Cabinet Room in the White House to consider the decision. At the meeting, the Secretary of Defense, Harold Brown, the Chairman of the Joint Chiefs of Staff, General David C. Jones and the National Security Adviser, Zbigniew Brzezinski, all spoke in favour of going ahead, and only the director of the CIA spoke against. President Carter went along with the majority. 'I feel comfortable with the decision', he said. In fact, though he had endorsed it, the decision had effectively been made beforehand, in the course of a struggle over the missile's design and specifications between the various agencies concerned. In a series of preliminary meetings, the Air Force, the Joint Chiefs of Staff, the National Security Adviser and Senator Henry Jackson had defeated the Secretary of Defense and the Undersecretary for Defense Research and Engineering. The real decision had taken place in the inter-agency struggle, with the President casting his vote with the winning side.

This decision was itself subsequently reversed when the Air Force failed to find an acceptable formula for mobile basing. The Reagan administration then reverted to siting the missiles in fixed silos, foregoing the mobility on which the rationale for the MX as a survivable missile had been based. Again the President was personally involved in the decision, this time taken in Reagan's sitting-room on the second floor of the White House. His three White House assistants, Ed Meese, James Baker and Michael Deaver were present, together with the Secretary of Defense, Caspar Weinberger and the National Security Assistant, Richard Allen. Once more, the decision had effectively been taken beforehand, since the Air Force and the Pentagon were determined to proceed with the missile and all of the mobile-basing options had been defeated or ruled out.

In the case of a technically, relatively trouble-free programme, such as the development of cruise missiles, there was little Presidential involvement, the key decisions being taken in the Office of the Secretary of Defense. As a side-effect of his decision to cancel the production of the B-1 bomber, President Carter ordered that the development of the air-launched cruise missile be accelerated. Later, cruise missiles were to become matters of Presidential concern in the context of arms control, but at the development stage the President was not otherwise involved.

Like the Presidency, the National Security Council has no regular role in nuclear weapons decisions, meeting only when conflicts between agencies need to be resolved. Its members are the President, Vice-President, the Secretaries of State and Defense, the director of the CIA and the Chairman of the Joint Chiefs of Staff. Several dozen NSC staff are based at the White House and provide support for the NSC's meetings. The National Security Adviser directs these staff and thus manages the flow of diplomatic and military information which passes through the White House. The National Security Adviser is a key post, and one that is not subject to Senate ratification.

Special *ad hoc* advisory commissions are frequently used to consider disputed strategic weapons issues and these too may play an important role. For example, when the Reagan Administration came into office, it set up the Scowcroft Commission to advise on the MX and the future development of US strategic forces. The Commission, whose members were picked by Weinberger, included Michael May, the former associate director of Livermore, General Bernard Schreiver who had directed the Minuteman missile programme, and other defence industry heads, former Pentagon officials and retired generals.

At the top level, policy on nuclear weapons in the USA is made, not by a single decision-maker, but rather as the outcome of power struggles between the Armed Services, the Joint Chiefs of Staff, the civilian officials in the Pentagon and the Secretary of Defense, with Congress an occasional player. It is quite possible for weapons to be produced or deployed which no-one wants, as a result of inter-agency disputes and compromises. The focus of decision-making constantly shifts from one level to another, with first one agency or institution, and then a different one, acting as the focus. The top-level decision-making is thus a mosaic of different actors who can sometimes be on the scene for a short time, blocking, facilitating, or modifying the course of development. It is through this confusing and shifting *mélange* of top-level bodies, many convened on an *ad-hoc* basis, in which personal relations and informal discussions have as much importance as formal decisions, that decisions are made. The lack of clarity of the lines of decision at the highest level is one of the factors which give so much power in the United States to those lower down the chain of authority.

The upper-level bodies may be likened to a series of doors through which a weapons programme has to pass. Intermittently, some of the doors may open and some may close. But if those who want the weapons keep pushing, in the end they will get through.

*   *   *

Every year on the anniversary of the Revolution, in a scene which has no parallels in the West, intercontinental missiles trundle over the cobbles in Red Square as part of a large military parade. The Square is bedecked with pictures of Lenin and the buildings of the Kremlin and the cupolas of St Basil's Church form an impressive backdrop. On the dais, members of the Politburo salute stiffly, wearing solemn expressions beneath their fur caps. For the Politburo, this parade is a celebration of the Party's accomplishments and of Soviet strength.

It is also a matter of personal pride for the members of the Politburo. For, to a far greater extent than their top-level counterparts in any Western country, members of the Politburo are personally responsible for driving nuclear weapons programmes forward. In the Western powers, by contrast, only a handful of senior ministers, always a very small minority of the Cabinet, take such decisions.

According to an authoritative Soviet work, 'military development is at the centre of attention of the Politburo of the Communist Party of the Soviet Union, which adopts resolutions concerning both the defence of the country as a whole and also concrete measures directed towards the improvement and development of the Armed Forces'.

The formal process of weapons procurement provides for the involvement of Politburo members at two key stages in every nuclear weapons programme: at the stage when a programme goes from an Advance Project to a Design (i.e. before the beginning of prototype development work) and at the decision to put a project into series production. The routine involvement of the Politburo in this way is quite unlike the role of any Western Cabinet.

In recent years the Politburo has included the Minister of Defence (a military man), the Party Secretary with responsibility for the defence industry (currently Zaikov) and the head of the KGB. Within the collective leadership, the General Secretary is the dominant figure. The Party leader is also the Chairman of the Praesidium of the Supreme Soviet (the top State body) and in effective command of the armed forces. The Prime Minister, who chairs the State Council of Ministers, is also a member of the Politburo.

The Politburo is elected by the Communist Party Central Committee, a body of some 320 full members and 150 candidate members. About 8 per cent are uniformed members of the armed services, and a further 4 per cent are defence industry ministers or scientists, designers and managers associated with the weapons industry.

Much of the staff work of the Politburo is carried out by the Central Committee Secretariat, which numbers about a thousand, organised in 23 departments. One of these deals with the defence industry. The Party Secretary in charge of this department is an important figure. When Brezhnev held this post in the late 1950s, his office is said to have been 'a kind of staff headquarters where the most important problems of missile technology were resolved, and meetings held with the participation of the most eminent scientists, designers and specialists in various fields of science, technology and production. L. I. Brezhnev was often seen in the factories where the missile technology was being created'.

Beneath the Politburo, the top-level body with responsibility for nuclear weapons is the Defence Council, the top national security agency of the Soviet Union. Its members are thought to include the Party General Secretary, the Minister of Defence, the Chairman of the Council of Ministers and the head of the KGB – all themselves Politburo members. The Defence Council co-ordinates the defence activities of the State, decides basic policy questions over the country's military development, and plays an important role in arms control decision-making. It probably oversees the weapons acquisition process, and would become the supreme organ of the State in time of war.

Another important body is the Military-Industrial Commission which is concerned with the co-ordination of the defence industry and the fulfilment of plans for the production and delivery of weapons to the armed forces. It considers the draft plans for the defence industry prepared by GOSPLAN, the State Planning Committee, and submits proposals on military industrial issues to the Defence Council. It is also thought to have responsibilities for co-ordinating weapons research and development and for co-ordinating the different defence industry ministries involved in weapons programmes.

All the main threads of information and decision-making concerning nuclear weapons come together in the Defence Council, the Defence Industry department of the Central Committee Secretariat and the Politburo. Outside ministers or officials are invited to join discussions as necessary. In the lower reaches of the Soviet bureaucracy there is rigid compartmentalisation, but at the top there is considerable informality and flexibility. 'The lines between the top leadership and the problems to be solved are short and direct', according to a staff member of the Soviet Institute of the USA and Canada.

Policy differences do arise in the Soviet government, and sometimes sharp changes of policy occur. There is little evidence, however, of strong disagreements on nuclear weapons policy. On the contrary, there has been a consensus in the Soviet Union on the necessity to maintain nuclear forces in parity with the United States. This consensus, together with the slow turnover in membership of the Politburo and its detailed oversight of nuclear weapons policies has lent coherence and continuity to Soviet nuclear policies.

*   *   *

Near the Forbidden City, off Tien An Min Square in Beijing, is the Zhongnanhai. The name means 'the central and southern seas', after the two lakes which are enclosed in the compound. Originally a playground for the Ming and Qing emperors, the site is now the headquarters of the Chinese Communist Party Central Committee. It is there, screened from the street by a tall wall, that the Chinese Politburo meets.

For many years a small group of top leaders who had known each other since the days of the Long March have taken the key decisions on China's nuclear weapons. In China, the major decisions on nuclear weapons have rested in the hands of less than fifty individuals.

In fact, the key body which has directed the nuclear weapons programme appears to be not the Politburo itself but the Central Military Commission, a top level committee which includes Politburo members and also meets in the Zhongnanhai. For long periods the membership and activities of the Central Military Commission have been concealed, but it appears to have exercised semi-autonomous control of military policy. This body may have provided continuity for the nuclear weapons programme at times when the Party leadership was in turmoil.

In 1985, Deng Xiaoping's chairmanship of the Central Military Commission was his highest official post. He used the position to cut back the People's Liberation Army by a million men, while continuing to modernise the strategic nuclear force.

Changes under Deng Xiaoping's leadership have begun to replace the old guard with a younger group of leaders on the Politburo, more representative of technocrats and managers than of the revolutionary leaders of the past. The influence of the military has also been reduced. At the same time the generation of pioneering nuclear scientists and rocket engineers who, together with the Party leaders, developed the first Chinese nuclear weapons, is giving way to a second generation which has grown up working in the nuclear programme.

These changes may bring about a change in the administration of nuclear weapons in China, which appears to have been in the hands of a group of leading scientists and engineers on the one hand, and top Politburo members and Party leaders on the other. If the nuclear force grows larger, there is likely to be a larger role for middle level managers and for the military.

Three elderly Marshals and veterans of the Long March survived Deng's recent changes and still sit on the Politburo in 1986. They are Yang Shungkun, who at 78 is only three years younger than Deng

Xiaoping himself, Yang Dezhi, and Yu Qili. Yang Shungkun is the vice-chairman of the Central Military Commission, Yang Dezhi is Chief of the General Staff of the People's Liberation Army and Yu Qili is the direction of the Army's Political Department and a member of the Central Military Commission. The continuity of the nuclear programme is being maintained.

*   *   *

The top-level policy-making body in French nuclear weapons decisions is the Conseil de Défense, chaired by the President of the Republic. Its meetings are held in secret. A meeting on 30 October 1982 took the decision to modernise French nuclear forces and approved a seventh nuclear submarine, the new tactical missile 'Hades', and the mobile strategic missile 'SX'. These decisions were made public two weeks after the meeting had been held. Mitterand explained to his stunned colleagues in the Socialist Party that he was acting on the basis of new and secret intelligence which had been made known to him when he came into office.

The President of the Republic is undoubtedly a key decision-maker, approving all the critical decisions on defence and nuclear weapons issues, on the basis of advice prepared within the government. He is also Chief of the French Armed Forces, and there is a Personal Military Staff in the Elysée, who carries the codes for launching French nuclear weapons. Other members of the Conseil de Défense include the Prime Minister, the Foreign Minister, the Defence Minister, the Finance Minister, the Interior Minister, the four Chiefs of Staff and the Chief of the President's Personal Military Staff. The secretary of this body is an official called the Secretary General for National Defence, who is also responsible for providing the President with intelligence reports.

The French Prime Minister is normally not an important decision-maker in nuclear weapons matters, as defence and foreign policy are traditionally prerogatives of the President. Nominally, the Prime Minister oversees the French atomic energy commission, the CEA, but this authority is usually delegated to another Minister and impinges little on the CEA's autonomy. The Prime Minister may sometimes take a role in presenting policy decisions to the general public. For example, Jacques Chirac, the Prime Minister in 1975, announced that the 'Pluton' short-range, tactical nuclear missiles had become operational. 'We have the technical, industrial and financial capacity to develop tactical nuclear weapons,' he said. 'It is logical that we seek to profit from this.'

*   *   *

Perhaps it is not surprising, considering the different social and political climates in the five nuclear states, that nuclear weapons are controlled at the top level in very different ways. In the Soviet Union and China, the Party leaderships, through the Politburos, have exercised close control over the development of nuclear weapons. The lines of authority are clear and unambiguous. The Politburos include people who have been concerned with nuclear weapons programmes over many years. In the Soviet Union, the Politburo is formally involved in key stages of each weapon's life-cycle, and no fundamental nuclear decision is taken without its consideration and approval.

In Britain and France, formal nuclear decisions are restricted to a narrow circle of top politicians, although, because of their relatively rapid turnover, their role in controlling weapons programmes is neither so continuous nor so central as that of the Politburo. In both Britain and France, the permanent bureaucracies have great importance in maintaining the continuity of decision-making and in shaping the options which are presented to the politicians.

In the United States, lines of authority and decision are neither so clear nor so centralised. Different top-level agencies, including special commissions, the National Security Council, the Cabinet and the President, intervene in the process, but their interventions are not necessarily final. The poitical leadership can initiate or block weapons projects, but the military and the bureaucracy can re-submit projects that have been cancelled after a change of Administration and modify or interpret projects that have been approved, effectively undoing top-level decisions.

In the West, those who administer nuclear weapons programmes have penetrated to high levels of government, where they have operated in great secrecy and are subject to erratic political control. In the East, the programmes are administered at the highest levels of government, and the continuity of leadership has been remarkable. In all five states, those who take the decisions to develop and deploy nuclear weapons are effectively insulated from public opposition and outside restraint.

7 Secrecy

[The H-bomb decision – the clearance system in Britain – classification systems in the United States – the KGB and Soviet security – secrecy and democracy]

In 1949 a secret internal debate took place in the USA among the President's advisers over whether or not to develop the H-bomb. Oppenheimer and many other scientists who had been involved in the A-bomb programme opposed it, since the H-bomb dwarfed even the A-bomb in destructiveness. The debate is described by Herbert York, an ex-director of the Livermore Laboratory, in his book *The Advisers: Oppenheimer, Teller and the Superbomb*: 'the argument was between moderates and hawks, or perhaps between hawks and superhawks. No full-fledged doves . . . were involved for the simple reason that none of them had the necessary clearances'. Security procedures designed to keep nuclear secrets had also served to restrict the circle of decision-makers to those who were prepared to accept and develop nuclear weapons.

Only about a hundred people in the USA knew of the meeting of the General Advisory Commission of the Atomic Energy Commission that was called to advise whether the US should embark on a crash programme to develop the thermonuclear bomb. The committee met under the chairmanship of Oppenheimer. Although this group favoured atomic weapons, they advised against the H-bomb: 'We all hope that by one means or another the development of these weapons can be avoided. We are all reluctant to see the United States take the initiative in precipitating this development. We are all agreed that it would be wrong at the present moment to commit ourselves to an all-out effort towards its development.' In a more strongly-worded minority report the scientists Rabi and Fermi wrote, 'the fact that no limits exist to the destructiveness of this weapon makes its very existence and the knowledge of its construction a danger to humanity as a whole . . . we believe it important for the President of the United States to tell the American public and the world that we think it wrong on fundamental, ethical principles to initiate the development of such a weapon'.

Despite this recommendation, the advocates of the H-bomb lobbied the Air Force, the Joint Congressional Committee on Atomic Energy,

Secretary of Defense Johnson, Paul Nitze, head of the planning division in the State Department, and Bradley, the Chairman of the Joint Chiefs of Staff. All of these came out in favour of development. Nitze believed it was essential for the world to go on believing in the superiority of American technology. The Chairman of the Joint Chiefs of Staff said that he could not bear to think that the Russians might be the first to produce the hydrogen bomb.

Suddenly, in 1950, as this lobbying was proceeding to the apex of the US government, Klaus Fuchs was arrested in England as an atomic spy. A deputy chief scientific officer at Harwell, he had been centrally involved in the British atomic research programme. The news that he had been passing atomic secrets to the Russians for years caused a sensation in both Britain and the United States.

The day after his arrest, a special committee of the National Security Council met to consider the H-bomb decision. The committee consisted of the Secretary of Defense, the Secretary of State and the Chairman of the Atomic Energy Commission. Impressed by the Fuchs case, they decided, by two votes (the Secretaries) to one (the Chairman) to recommend to the President that a crash programme should proceed.

The same afternoon, President Truman announced to the American people, 'I have directed the Atomic Energy Commission to continue its work on all forms of atomic weapons, including the 'hydrogen' or super bomb. Like all other work in the field of atomic weapons, it is being and will be carried forward on a basis consistent with the overall objectives of our programme for peace and security.'

*   *   *

In the context of nuclear weapons, the word 'security' has acquired twin meanings. It refers to safety, but also to secrecy and secure communications. In the war, secrecy of atomic information was considered essential to prevent the Germans from learning of the US programme to make the bomb. After the war, the authorities were equally concerned to prevent information reaching the Russians. In both the US and Britain, therefore, powerful measures were taken to ensure security. These measures still influence how nuclear decisions are taken and how nuclear weapons issues can be discussed.

In Britain the fundamental measure is the Official Secrets Act. The first task of every civil servant is to read this Act, which makes it a criminal offence for any servant of the Crown to communicate or publish any official information without authority. The Act covers not only classified information but any official document. It has been used most recently against Sarah Tisdall, a clerk in the Foreign Office, who sent copies of two documents about the arrangements for deploying cruise missiles at Greenham Common to the *Guardian* newspaper and in the unsuccessful prosecution of Clive Ponting, who passed documents, indicating that the government was misleading a Select Committee, to an MP.

The Official Secrets Act makes it a crime to receive any unauthorised official information, unless the recipient proves 'that the communication to him . . . was contrary to his desire'. Its penalties thus fall on the press as well as the civil servant. The press is also prevented from disclosing information about defence and intelligence matters by the D-Notice system. D-Notices have no legal force but are widely respected by the media. The D-Notice Committee is composed of defence and intelligence officials and newspaper editors. Together they advise on material which should not be published. D-Notices have included, for example, details of transfers of nuclear materials, of the stockpile of nuclear weapons, and aspects of the production of nuclear weapons. The nature and purpose of the Atomic Weapons Research Establishment at Aldermaston was kept secret for several years under this system.

Within the government, the classification of documents serves to control the flow of information on sensitive matters. There are four levels of secrecy. These are 'Restricted', 'Confidential', 'Secret' and 'Top Secret'. 'Top Secret' documents are further sub-divided into more restricted classifications such as 'Top Secret (Atomic)' and top secret intelligence material. Beyond Top Secret (Atomic) there are yet further classifications for the nuclear design secrets which even senior Ministers are not allowed to see. No junior Ministers are cleared to see any Top Secret information. This is generally restricted to Privy Councillors, who are bound by an Oath of Confidentiality which dates back to 1250.

For civil servants, who are not bound by the sanction of a royal oath, access to secret documents is controlled by a vetting system. All civil servants at the Ministry of Defence receive negative vetting, which is a check through the computer records of MI5 (the counter-intelligence agency) and the Criminal Records Office. In order to see 'Top Secret' documents it is necessary to be positively vetted. This entails inquiries into the past life and acquaintances of the subject and is intended to reveal any past or present affiliations which Communist or Fascist organisations which might raise doubts about reliability, and any pecuniary embarrassment, sexual liaisons or serious moral or habitual weaknesses which might make the subject a security risk. All civil servants in the 'fast stream' of those eligible for top posts in the MoD are positively vetted. Altogether some 16 300 military posts and 11 700 civilian posts, including posts in the Ministry of Defence, Aldermaston and the Government Communications Headquarters (GCHQ) are subject to positive vetting. As a further security measure, lie-detectors and psychological tests have been introduced.

A number of top-level bodies control the implementation of these measures. Within the Cabinet Office there are Security and Personnel Security committees, chaired by the Cabinet Secretary; several cabinet subcommittees also monitor security matters.

The security system for nuclear information automatically restricts the number of people who can participate in the taking of decisions in this area. Therefore outsiders, even those who are well-informed and well-disposed towards the defence community, are considered to be unable to contribute to the decisions. These outsiders, who by definition lack the really 'hard' information, include not only ordinary members of the public, academics, and journalists but also Members of Parliament and the leading members of the Opposition.

Most Cabinet Ministers are also excluded from the sensitive nuclear papers. Indeed, the inner circle of Ministers who have had access to these papers have believed even more fervently than civil servants in the importance of maintaining their secrecy. Attlee was loath to release any atomic information, even when on occasion it was already publicly available in some form. His inner Cabinet's decision to develop nuclear weapons was not transmitted even to the full Cabinet, let alone to Parliament and the public.

Therefore, the decision by Britain to acquire nuclear weapons was made without parliamentary approval or public consent. Churchill is the only Prime Minister to have consulted the full Cabinet on a nuclear weapons question. There were three full discussions in 1954 on the decision to build a British H-bomb. Parliament, however, was not informed until 12 months later. Information about Chevaline was withheld from Parliament for 13 years, and all the preparatory studies leading to the Trident decision were conducted in secrecy.

Certain details of the deterrent policy are kept even more secret than procurement decisions. 'Though I have been the minister responsible for the Atomic Research Centre at Aldermaston', wrote

Tony Benn, 'and have served in four cabinets and on occasions as a member of the key Overseas Policy and Defence Committee, as well as other, smaller committees dealing with nuclear policy, I was never told, and still do not know, the basis upon which US nuclear weapons sited in the UK can be fired. The general assumption is that guidelines have been drawn up that constitute a working agreement governing their use and provide for consultation between President and Prime Minister, if that is practical. No Cabinet in which I have served has ever been told the true position and I can only suppose that the key US/UK arrangements are, in effect, only known to the President and the Prime Minister.'

According to a frank careers guide prepared by the First Division Association (which represents senior civil servants), 'Her Majesty's Government runs one of the most secretive democracies in the world.' In no area is the government more secretive than in matters concerning nuclear weapons.

*   *   *

In Britain the secrecy system works effectively because those who are involved in nuclear decisions form a narrow élite within the government. In the United States, the system for classifying documents is even more elaborate than in Britain, but so many people have access to secret information that access to nuclear information is relatively open compared with other nuclear countries.

There are at least 47 different levels of classification, with each level restricted to a smaller number of people than the one below it. The most secret information is that covering nuclear warheads. This is covered by the 'Q-clearance' system, administered by the Department of Energy. Beyond the 'Q-clearance' there are further levels of classification, but the existence of these levels is itself classified.

At the lowest levels of classification, secret information is very widespread. Over four million Americans – out of a population of some 200 million – are cleared to see secret information of some kind. This extraordinary number means not only that much classified information is not secret in any meaningful sense, but also that the culture of secrecy has extended outside the government to reach a sizable section of American society.

This extensive dissemination of classified information is explained, in part, by the fact that little of it is classified to protect information from foreign powers. Rather, documents are classified because it is in the interests of the department that originates them to do so. Commenting on how classification is used by the Navy, retired Admiral Gene LaRocque said that 'about 50 per cent of what is classified in the Navy is designed to keep it from the Army and the Air Force and the Secretary of Defense. About 25 per cent is designed to be kept secret from Congress and about 20 per cent from the State Department and the public'. He concluded that only 5 per cent of all secrecy in the Navy was justified by security regulations.

Secrecy also preserves the interests of the defence contractors, whose pursuit of new weapons designs and contracts can be carried out among a community of people who share the same privileged information, and are unencumbered by the need to justify their plans in an open forum.

Because the decision-making system in the US is diffuse, and people from the defence contractors, the weapons laboratories, the Pentagon and the government contribute to the making of nuclear weapons decisions, nuclear information, though secret, is widespread. It is also less tightly restricted than in other countries. According to Dr Jack Ruina, former head of the Defense Advanced Research Projects Agency (DARPA), nothing of any technical significance remains secret in the USA for more than a few days. Even news of the outcome of tests on the X-ray laser, for example, became publicly known within days, although the very existence of the project was supposed to be secret. Secrecy is broken often by those who wish a weapon to get adopted. Publicity can be helpful sometimes in winning support.

The US Constitution enshrines the freedom of information in the first Amendment, which prohibits Congress from passing any law limiting freedom of speech. In 1966 Congress passed the Freedom of Information Act, amended in 1974, which gave a public right of access to government information. However, despite these important checks on secrecy, a series of legal cases have established that the government has a right to prevent publication of material which affects national security.

The amount of information that is publicly available on nuclear weapons in the USA vastly exceeds what is available in the other major nuclear countries. Senator Thomas McIntyre, while he was chairman of the Senate Armed Services Subcommittee on Research and Development, complained of the information overload his committee faced: 'we spend an awful lot of time, but we are lucky if we can take a look or have a briefing or hearing on, say, 15 per cent of those projects (submitted to the committee for approval)'.

Nevertheless, the secrecy system does have importance. It serves to draw a line between those who have privileged information and the wider public who do not. In so doing, it effectively discriminates between those who can and cannot contribute to the internal policy discussion over nuclear weapons.

Secret information is regarded as important simply because it is secret. Those who have access to it tend to regard themselves as important for that reason. It means they are trusted and powerful. Those with access to secret information form, in a sense, a secret society. Those who lack this information are regarded as not fully in the picture, and hence unable to contribute fully to decisions. When Oppenheimer lost his security clearances, he regarded it as a bitter blow, as it meant he could no longer contribute to nuclear policy. The effect of the secrecy system is to exclude those with inadequate clearances from the decision-making system. It acts as a filter to ensure that those who have access to the highest levels of information are reliable and have acceptable political views. Moreover, those who have been cleared to receive high levels of secret information feel unwilling to hold discussions on matters related to nuclear weapons issues with persons not so cleared, lest they give away information which is secret and thereby commit an offence. Not surprisingly, this makes those who are involved in nuclear work guarded and secretive by nature, and any form of dialogue with outsiders becomes prejudiced.

*   *   *

If official secrecy is such as to curtail democratic decision-making in the area of nuclear weapons in the West, in the East, where the government cannot be dismissed by the electorate in elections, the availability of information is very much lower. In the West, details of decisions and crucial meetings often leak out after the event. In the Soviet Union, even retrospective information is not published about recent decisions.

This undoubtedly owes a good deal to the way in which the nuclear weapons programme was organised when it began. During the Stalin era, the atomic plants and the design bureaux were under the authority of the secret police, under Beria's overall charge. When Beria fell, these arrangements were changed, but the KGB retained responsibility for the security of military R & D and for the security of communications between institutions involved in nuclear weapons production and design. Each such institution has a KGB division which checks on the reliability of employees and controls access to classified information.

Even within the government, information is tightly controlled and highly compartmentalised, so that only those at the top have a full picture of what is going on. Moreover, different parts of the government attempt to keep information from one another, often successfully. The General Staff is unwilling to divulge military information to other civilian departments, and Western observers at arms control negotiations have been surprised to notice that civilian negotiators have had to be corrected or informed by their military colleagues as to details of what weapons the Soviet Union deploys. Similarly, the defence industry keeps secret details of production, in part to avoid criticism of its efficiency. Khrushchev noted, 'because the production of defence industry enterprises is secret, shortcomings in the work of such enterprises are closed to criticism . . . The defence industry is coping successfully with creating and producing modern weapons. But these tasks could have been carried out more successfully and at a lower cost'. Only the Politburo, the Defence Council, the Defence Industry Department of the Central Committee Secretariat and probably the Military Industrial Commission, the top organs of Soviet decision-making, have access to full information.

The Soviet public is denied any information beyond what the government wishes to disseminate. The government's control of the media and the absence of an independent, critical press contributes to the lack of open debate on nuclear weapons issues.

This intense degree of secrecy originates in and reflects the divisions within Soviet society, between the governors and the governed. Tight control of information ensures that only those at the top are able to contribute to decision-making on national security issues, and vests in the top leadership of the Party centralised control of nuclear weapons policy.

*   *   *

In all five of the major nuclear states, nuclear affairs have become high secrets of state and the classification system has been used to restrict those who have a say in them. This has had important consequences.

In the West, nuclear decisions are taken behind closed doors and this gravely restricts accountability to the public. Ministers cannot be called to account for decisions which are not known. In Britain, for example, only one person in the Ministry of Defence can be accountable for decisions on nuclear weapons – the Secretary of State for Defence. The junior Ministers cannot be fully accountable since they are not cleared to receive the top nuclear papers.

Secrecy has permitted the nuclear weapons laboratories and the defence contractors to develop new weapons systems according to their own criteria, without public scrutiny or criticism. At the early stages of decisions to formulate a new nuclear project or start work on a weapons system, information is withheld from the public, yet this may be the critical phase in the decision.

Secrecy has drawn a barrier around those who are involved in the 'weapons' or nuclear weapons 'community' and those who are excluded from it. It has served to maintain a sharp line between those who accept the need for developing new nuclear weapons, which is effectively a requirement for high-level clearances, and those who are critical of new nuclear weapons, who by their lack of clearances are excluded from policy-making.

Those who have access to secret information become thereby privileged, and unconsciously accept the rules of the club they have joined. In order to get more information, it is necessary not to rock the boat too much. Consequently, the subcommittees of Congress and the Select Committees of Parliament, to the extent that they receive secret information, become privileged and become advisers. The effect is to mute criticism and to encourage acceptance of secret decision-making.

Between the nuclear countries, the existence of secrecy in research, development and production, especially in the Eastern countries, has encouraged exaggeration of capabilities. In the Soviet Union secrecy may have been intended to conceal inferiority, but now its retention has a destabilising effect. Not only does it encourage worst-case assessments, it also allows defence contractors, nuclear weapons scientists and military circles in the West to claim Soviet advances as a justification for their own new weapons projects. The development of MIRVed missiles, the neutron bomb and the Strategic Defence Initiative were all justified in these terms in the USA. Since the intelligence estimates on which such claims are made are themselves secret, the public has no way of assessing these claims.

Nuclear weapons have bred secrecy and secrecy has sustained nuclear weapons. By allowing the growth of secret government in this area, the public has lost its right to know whether, how and under what conditions its government will go to war.

8 Parliaments

[The role of parliaments and representative assemblies in nuclear weapons decisions: the House of Commons – the French Parlement – the Supreme Soviet – the National People's Congress of China – the US Congress – the limits of parliamentary control]

From Waterloo Bridge at night, the centre of London is a majestic panorama of flood-lit buildings. St Paul's Cathedral stands out amongst the modern commercial towers to the east, bathed in yellow light. To the west, the Houses of Parliament present their well-known façade to the river, bronze and stately. A little way along the river, to the east, stands the Ministry of Defence, with flags on its roof and lights ablaze in its office windows. Although only a few hundred yards separate the Houses of Parliament and the Ministry of Defence, decisions about nuclear weapons can be taken within the Ministry without Parliament's knowledge or approval.

On the 9th December 1981, Sir Frank Cooper, then Permanent Under Secretary at the Ministry of Defence, was giving evidence to the House of Commons Committee of Public Accounts. He sat in the oak-panelled committee room, its walls hung with sombre oil paintings of distinguished British public figures of the 18th and 19th centuries. Behind him was a small phalanx of civil servants. In front, he faced nine members of the parliamentary committee, with a Labour ex-Cabinet Minister, the Rt. Hon. Joel Barnett MP, in the Chair. Barnett was asking Sir Frank Cooper to explain how it was that parliament had not been told about the Chevaline project for 14 years.

Chairman: On the question of accountability, in paragraphs 12 and 13 of your paper the costs of Chevaline are not separately identifiable within the separate votes involved, namely Vote 2. So Parliament is not informed by the Supply Estimates of the total funds to be spent on the project in any year, nor are we provided with a total cost to date and to completion of the project. In fact it was not until January 1980 that the House of Commons was informed by the Secretary of State that the Chevaline programme was estimated to cost £1,000 million. Could you tell us why this is and whether you would regard the situation as satisfactory?
Sir Frank Cooper: I agree that it was not until January 1980 that the

full cost was disclosed. If one goes back, there is a whole series of questions and answers which dealt in part with this. If I go back to 5 February 1973 Mr Allaun was asking, "Is it correct that the Government have spent £100 million on updating these missiles?" To which he got the answer, "We have no intention of abandoning our nuclear capability. It forms a valuable contribution to NATO deterrent forces." There is a whole series of answers from December 1974 onwards about maintaining the effectiveness of our capability.

Chairman: It is the answers more than the question with which I am concerned. The answers are not really dealing with the point I was making, are they?

Sir Frank Cooper: No, they are not, but I think there is a whole series of parliamentary exchanges about what became known as Chevaline.

Chairman: Do you consider that true accountability to parliament or anything remotely like it?

Sir Frank Cooper: I am not a member of Parliament.

Chairman: But we are.

Sir Frank Cooper: The tradition in Parliament is that you ask Ministers about this kind of thing.

Chairman: Not in this Committee it is not. We ask you.

Sir Frank Cooper: Well, could I say this. What I do believe is that it would be wrong, for quite a period of time, to disclose in public the nature of the Chevaline project or its costs. It was necessary for a period of time for that not to be brought into the public gaze.

The view that it would be wrong to inform the public of the nature of a nuclear weapon project in its early stages has not been exceptional among those who make decisions on nuclear weapons. At the outset of the nuclear era, Attlee's Government took the decision to withhold information about the atomic programme from Parliament and to conceal the estimates for expenditure on atomic weapons by burying them under a subheading called 'Public Buildings in Great Britain' in the Civil Contingencies Fund of the Ministry of Supply vote.

From 1965 to 1979 no major debates were held on nuclear weapons, apart from the general debates on the Defence Estimates which take place annually. During this period, decisions on Chevaline by the Wilson, Heath and Callaghan governments were not revealed to parliament. Wilson later justified his decision to restrict discussion of Chevaline to a small group of Ministers in the following terms: 'It isn't a question of not trusting. It's a question that the more people you

have, the more people can be got at, for example by backbenchers who
then start to press Cabinet Ministers.' No information was provided to
Parliament about Trident during the critical period between 1976 and
November 1979, while internal government discussions and secret
negotiations with the Americans were underway.

Since 1979, when the Thatcher government came into office with a
Labour Opposition for the first time committed to unilateral disarmament, the amount of discussion in Parliament of nuclear weapons has
substantially increased. The decision to acquire Trident, and later to
acquire the more expensive Trident II, were followed by debates, in
which the Government's policy was endorsed by substantial majorities. There was also a debate on Cruise shortly after it was deployed, in
which the Government won a large majority on a motion reaffirming
the support of the House of Commons for the 1979 NATO twin-track
decision.

These long, set-piece debates in the House of Commons provide an
opportunity for most shades of opinion to be aired. In addition, there
are annual debates on defence and on government public expenditure
plans, in which nuclear weapons can be discussed. In the votes which
follow these debates Parliament has an opportunity to vote for or
against the government. However, no government has ever lost a
motion on nuclear weapons policy, and unless there were to be a hung
Parliament, it is most unlikely that this could occur. Strong party
discipline, the two-party system and three line whips normally enable
the government in power to muster a majority in the House of
Commons.

Moreover, debates on nuclear weapons issues are held after, not
before, the decisions on these issues have been taken by the small
circle of senior civil servants, military officers and Ministers of the
inner Cabinet. Similarly, decisions taken by NATO Ministers may
commit the government to a position before Parliament has been
consulted. The Montebello decision to modernise theatre nuclear
forces and the NATO decision to adopt a strategy of 'Follow-On Force
Attack' (that is, a strategy envisaging deep strikes behind Warsaw Pact
lines in the event of war) were decisions of this kind. Parliament's role
is therefore limited to endorsing decisions already taken. While this is
not unique to nuclear weapons issues, it means that Parliament does
not participate in the decision-making process.

If Parliament were to exercise control over nuclear weapons before
the decisions were taken, this control should come through its control
of expenditure. All the money which is spent on nuclear weapons is

voted through by Parliament. However, at present this expenditure is accumulated into a single heading in the Defence Estimates, the 'Nuclear Strategic Force'. Parliament is not told how this is broken down, line by line, into particular weapons systems and items of equipment. Consequently, Parliament's financial control of nuclear weapons has been effectively delegated to those in the government who hold this information.

The Parliamentary body which examines government expenditure the most rigorously is the Public Accounts Committee. This Committee examines how money has been spent in particular programmes and can take evidence from civil servants, Ministers and others. Its powers are limited to retrospective inquiries. When it became clear that expenditure on Chevaline had not been disclosed to Parliament for 14 years, the Public Accounts Committee wrote:

> Expenditure each year was included in the normal way in the Defence Estimates and Appropriation Accounts; our criticism is that the costs were not disclosed and that there was no requirement that they should be disclosed. Incidental and oblique references to a Polaris enhancement programme made in Parliament or to Parliamentary committees in our view do not provide sufficient information for Parliament to discharge its responsibility to scrutinise major expenditure proposals and to exercise proper financial control over supply . . .
> . . . The failure to inform Parliament or this committee until 1980 that a major programme on this scale was being undertaken, or that its cost was turning out to be so far in excess of that originally expected, is quite unacceptable. Full accountability to Parliament in future is imperative.

After this report the government agreed to provide the Parliamentary Accounts Committee with information about expenditure on major defence programmes costing more than £200 million. The Public Accounts Committee declared itself satisfied with this improvement. However in 1985 there were further clashes between the Public Accounts Committee and the Government when the Comptroller and Auditor General, who reports to the Committee, was denied two reports about waste in government spending. Since the Public Accounts Committee is primarily concerned with the efficiency and economy of government spending and not with whether particular programmes should be financed in the first place, it is not a particularly appropriate vehicle for exercising financial control of defence projects. The information that is provided to the Public Accounts Committee by the Ministry of Defence is secret and retrospective, coming over a year after the expenditure has been incurred. The Committee's only real power is that of rebuke.

The Comptroller and Auditor General said in a lecture at the London School of Economics in 1986: 'The answer to the simple, stark question: "Does Parliament get from the Government all the information it needs for effective accountability?" must today be "No".'

Besides the Public Accounts Committee, there is a House of Commons Select Committee on Defence. This Committee has undertaken several inquiries relevant to nuclear weapons. In 1980 it embarked on an inquiry into the question of replacements for the Polaris force and began taking evidence. Suddenly, before it had completed its inquiry, the government announced its decision on Trident. The Labour members of the committee complained in a minority report:

> Parliament's role in the decision to procure a successor system to Polaris has been limited to endorsing a decision already taken. Decisions on defence, and on Britain's strategic nuclear deterrent have historically been taken by a small élite of very senior Cabinet Ministers, civil servants and Service Chiefs, and this present decision was certainly no exception . . .
>
> The Government came to the House and invited it to endorse the Trident decision when the Committee was still deliberating. We saw no reason for action by the House before the Committee reported, and consider the Government's actions in this respect to be less than courteous to the House and its Committee.

The Select Committee has no power to ensure that it will not be treated in the same manner again.

Both the Public Accounts Committee and the Select Committee on Defence normally have a Chairman and a majority of MPs representing the party in government. They tend to split along party lines on nuclear weapons issues.

These parliamentary committees are unique as a means of wringing information from the government and as a forum for detailed cross-examination of civil servants and Ministers by MPs. However the committees have no powers and no teeth other than the influence of their reports.

Until 1979, neither of the major parties had a strong interest in bringing nuclear weapons to a debate. The Conservative Party supported the maintenance of the nuclear force and the Labour Party was divided. After 1979, the Labour Party changed from a policy of supporting the continuation of existing nuclear weapons to one of opposition to an independent British nuclear force and to the basing of US nuclear weapons in Britain. Nevertheless, it remained difficult for Members of Parliament to obtain information.

A good example is the story of the W-82 155 mm artillery-fired battlefield nuclear weapon. This is planned for deployment in Europe by the 1990s. The plans came to light in May 1984, when Dr Richard Wagner, Assistant to the Secretary of Defense for Atomic Energy, gave details of several new short-range nuclear weapons to a US Senate subcommittee. He said that NATO Ministers had endorsed the development of these weapons, including the artillery-fired nuclear shell. Subsequently Congress voted funds for the production of these weapons. In October 1984 an MP in the House of Commons asked the Secretary of State for Defence what his policy was towards the deployment of the 155 mm nuclear projectile. He was told, 'No proposals to this effect have been made.' In February 1985, another MP raised the matter again and drew attention to the discrepancy between Dr Wagner's statements to the Congress and the statements that had been given in the House. He was told by a Ministry that there were no specific NATO proposals. In March 1985 the Minister of Defence was again asked and again denied that any specific proposals had been made. In the same month the NATO Nuclear Planning Group announced in an unclassified communiqué that it had been agreed to modernise 'our forces across the spectrum of capabilities'. In April 1985, Dr Wagner was questioned again in the US Congress. It was suggested to him, on the basis of the replies in the House of Commons, that 'the European agreement is far less tangible and more tenuous . . . than we have been told'. Dr Wagner disagreed. 'I am rather certain that they have committed to the modernisation of the short-range systems. They discussed that', he said. He was asked: 'Is their desire not to be explicit for home consumption?' He replied, 'Probably.' In May 1985, Mr John Stanley, the Minister for the Armed Forces, was asked by an MP why there was no reference to the new nuclear artillery shells in the Defence Estimates. He replied that the details of the plans to modernise were classified and that 'no decisions have been made'. In July 1985, Mr Heseltine, the Secretary of State for Defence, said in a letter to an MP that no timetable for the modernisation decisions had been agreed. He said he would 'report to Parliament in respect of modernisation affecting British forces when the appropriate moment comes'. In November 1985, the Labour Defence spokesman, Mr Denzil Davies, called for a debate in Parliament on the modernisation of battlefield nuclear weapons before the decision was made. 'We are worried that at the end of the day we might have more powerful battlefield nuclear weapons in central Europe . . . such matters should be debated in the House of Commons before decisions are taken.' Mr Heseltine, replied: 'The House must think it extraordinary that a representative of a Government who modernised our independent nuclear deterrent by the Chevaline process without telling anyone they had done it should expect us to subject modernisation programmes for nuclear weapons to debate in the House. It is unthinkable.'

If a majority of MPs were elected who held different views to those of the present nuclear weapons decison-makers, if there were to be a hung Parliament, or if the powers of the Select Committees and the amount of financial information available were to be increased, it is conceivable that Parliament could play a greater role. It has the ultimate power to refuse to endorse government expenditure. But, in the British system, it is difficult to imagine Parliament exercising such powers in opposition to the government. Its main political importance, in the present climate, comes when it is dissolved. A new Parliament with a different composition of parties can then bring a new permutation of senior Cabinet Ministers into power.

While senior civil servants continue to hold the attitude that it is wrong to inform the public of nuclear weapons decisions in their early stages, and Ministers preserve the tradition of taking decisions in secret Cabinet subcommittees, there is little guarantee that a British government could not, once again, embark on a new nuclear or strategic project without the approval of Parliament or the knowledge of the public.

*   *   *

'There are military secrets concealed behind the silence of the budget.' So said Pierre Messmer, de Gaulle's Minister of Defence and later Prime Minister. In France, as in Britain, nuclear weapons items are frequently hidden in budget aggregates, which makes it difficult to find out how much is being spent, and on what. Although the French

Parlement votes on the defence budget, it has never exercised control of the nuclear weapons programme. There has been little debate in the Parlement of nuclear weapons. This has been partly because the government sets the agenda of the Parlement and partly because of the consensus between the major parties in favour of the nuclear force.

The only occasion on which Parlement voted against the government over a defence budget was in 1983, when the socialist government of François Mitterand proposed a cut in planned defence expenditure. There were storms of protest from military circles and the parties of the Right, and the Senate threw out the government defence budget. President Mitterand then presented it once more to the Assemblée, where there was a government majority, and it was enacted. Only if there was a majority opposed to the government in both assemblies could the government be defeated. As in Britain, the real importance of Parlement is at elections, when the government can be changed by the electorate.

*　　*　　*

The nearest equivalent to a Parliament in the Soviet Union is the Supreme Soviet. This is the highest legislative body of the USSR. Its 750 members are elected, with one deputy for every 300 000 of the Soviet Union's 270 million citizens. Unlike Western parliaments, however, this body meets only twice a year, and its functions are limited to enacting laws, approving state plans, and electing the Council of Ministers (the top body of the State government). It has several Standing Commissions which can examine the work of government bodies and initiate legislation, including a Commission for Foreign Affairs, but it has no Standing Commission on Defence and no role in defence matters. However, it does elect from among its members the Praesidium of the Supreme Soviet, a body of about forty members, which in turn forms the Council of Defence and has the constitutional power to appoint and dismiss the high command of the Armed Forces. The Supreme Soviet also forms the Council of Ministers, which includes the Defence Industry Ministers.

These elective powers are limited by the dominance of the Communist Party. According to the constitution, the Party is the 'leading and guiding force of Soviet society and the nucleus of its political system'. 'The Communist Party directs the activities of the Supreme Soviet and all Soviet workers' deputies. The decisions of

Party Congresses and the Central Committee are the ideological-political bases of laws and other acts approved by the Supreme Soviet of the USSR.'

The Communist Party consists of a hierarchy of bodies, with the Central Committee of the All-Union Communist Party of the Soviet Union at the top, a Central Committee for each Union Republic, and a range of intermediate organisations down to District and Town Party Committees. One in nine of all adult citizens are members of the Party. Power flows from the top down. The USSR Party bodies can overturn decisions made by the Union bodies, and higher levels of the Party instruct lower levels on how to implement the policies the Party has decided. Members of higher Party bodies are elected by lower ones, but the choice of secretaries by rural, urban and local district committees are confirmed by the next highest Party organisation. In principle, Party members have the right to make proposals about the 'policy and practical activity' of the Party, but in practice, official Party policy is rarely questioned or criticised.

A key feature of Soviet government is the system of 'nomenklatura', or nominated ones. These are 'leading cadres' selected by the Central Committee's secretariat for the top jobs. Similar perhaps to the 'fast stream' in the British Civil Service, it represents both a means of identification and promotion of the most promising candidates, and a mechanism to ensure that those who are recruited to the top posts conform to the Party orthodoxy. At the bottom of the hierarchy, new recruits to the Party are also chosen selectively by existing Party members.

The Party is the main vehicle for the change of policy, but because it is controlled from the top, change normally comes at the top. Major changes tend to occur with a change of leadership, when new leaders can introduce new policies and change the occupants of many posts. According to a Soviet law the Party has a 'monolithic unity in ideas and in organisation' and prohibits factions. In practice, there are different groupings and tendencies within the party which hold varied opinions on questions such as detente and priorities between defence and consumer spending.

The ordinary Soviet citizen is provided with little information or detail about Soviet nuclear weapons and there is little evidence that this is an important political issue. Although the Party might respond to a major shift in public opinion, there is little scope for those concerned with nuclear weapons either to oppose Soviet nuclear weapons policy or to become well informed about Soviet nuclear decisions. The accountability for nuclear weapons decisions to ordinary Soviet citizens is thus even more limited than it is in Britain and France.

* * *

Like the Soviet Union, China has no precise equivalent of the Western parliaments, and there are twin lines of authority through the State and the Party. The nearest equivalent to a Parliamentary institution is the national People's Congress, theoretically the highest authority of the State government.

The National People's Congress is elected once every five years, and is supposed to meet annually, although at times of political turmoil several years have gone by without the NPC being called. The delegates are elected indirectly by provincial congresses and there are reserved seats for members of the People's Liberation Army and the national minorities. The National People's Congress approves the economic plan, the budget and the state accounts, and has the power to amend the constitution. It also elects the President of the Republic, the Standing Committee of the NPC (which has legislative powers) and the chairman of the Central Military Commission – however there are no alternative candidates to choose from. Essentially the NPC is a means for elected representatives of the people to endorse State policy.

The sixth National People's Congress opened in the Great Hall of the People in Beijing on 6 June 1983. Two thousand nine hundred and seventy seven delegates were present, seated in semi-circular tiers in the large auditorium. Facing them were several rows of dignitaries and speakers on a stage, beneath a large red banner inscribed with Chinese characters.

Although the National People's Congress is in session only for a short time in the year, it does represent a forum in which debate can take place and public opinion can be expressed. People can write to the National People's Congress with complaints or suggestions, and some of these are selected for action to be taken. At the sixth Congress, sharp differences of opinion were expressed about economic strategy. Nuclear weapons, however, were not discussed. There is relatively little information about nuclear decisions in the Chinese media and China's nuclear weapons are not regarded as a contentious issue by the overwhelming majority of Chinese.

As in the Soviet Union, guidance for the State government comes from the Communist Party. According to the 1977 constitution, 'the party is organized on the principle of democratic centralism'. This means that 'the individual is subordinated to the organization, the minority is subordinate to the majority, the lower level is subordinate to the higher level, and the entire party is subordinate to the Central Committee'.

*   *   *

Washington – A House-Senate conference committee has approved a compromise military budget that Congressional aides said would restore money for all of the 22 weapons systems that either the House or the Senate had voted to kill.

The $302.5 billion military programs bill . . . assures that none of the weapons programs the Pentagon requested will be eliminated next year.

A source involved in the process said the bill for the 1986 fiscal year "proved once again that Congress can't kill weapons systems, any more than the Pentagon can."

(*International Herald Tribune*, 27.7.85)

Every US nuclear delivery vehicle and every US nuclear warhead has to be approved by the US Congress, not once, but twice, before funds can be legally transferred to their manufacturers. The programmes have to be approved by both Houses of Congress, the 435-strong House of Representatives, whose members have to stand for re-election every two years, and the 100-strong Senate, whose members face elections every six years. Both Houses are in session for most of the year.

In order to consider defence expenditure, the Congress has four full-time committees – the Armed Services Committees and Appropriations Committees of the House of Representatives and the Senate. Congressional approval is required both for the authorisation of weapons programmes and for the appropriation of funding for them. These committees themselves have subcommittees to which detailed matters are delegated. These four committees continuously review the military budget throughout the year.

The proceedings of Congress and its subcommittees are televised and published. The members of Congress are deluged by letters, telephone calls, computerised mailshots, public interest lobbies, corporate lobbies, and information by the sackful. Around an orbital expressway which circles the perimeter of the city are the offices of the 'Beltway Bandits' – defence contractors whose job it is to lobby the Government and the Congress on behalf of their companies. For, unlike the representative institutions in the other four nations, Congress has a real role in nuclear weapons decision-making. It is the forum in which the battle between the contending weapons programmes, corporations, and Service bids is fought out.

The Congressional review process generates an enormous amount of publicly accessible information about defence and about nuclear weapons – far more than is made available in any or all of the other nuclear states. In 1983 for example, the two Armed Services Committees published 22 000 pages of hearings and evidence. The Congressional budget review procedure enables the House and Senate committees to inspect defence spending on a line-by-line basis. Individual items are not aggregated and hidden as they are in the UK Defence Estimates. Moreover, a crucial difference between Congress and the French and UK parliamentary systems is that the Congress has the power to review government programmes before it authorises funding for them.

Congress has the power to increase or decrease the government's military budget, and it can endorse, modify or cancel funding for individual programmes. Even if funding is withheld, however, a new request can be submitted the following year. In practice, Congress has been reluctant to cancel major nuclear programmes and no major nuclear weapons system has been stopped outright. Many have been delayed or modified. For example, Congressional objections in 1973 to the Submarine-Launched Cruise Missile, on the grounds that its strategic rationale was unclear, led to a sharp cut-back in development funding. But in later years funding was restored.

However, it is not necessary for Congress to cancel programmes in order to exert influence. If Congress is likely to oppose a particular decision or programme, this usually becomes clear well before the request for appropriation comes before it, and the Pentagon or the President will normally withdraw or modify a programme from the budget rather than have it fall. It was Congressional opposition to the 'racetrack' scheme for basing the MX in the deserts of Utah and Nevada, reflecting the opposition of local people, which blocked the Air Force mobile-basing plan, although technical doubts and adverse reports by a special panel were also factors. President Reagan and his aides, together with the top officials in the Department of Defense, proposed instead the 'compromise' of basing the MX in Minuteman silos. Even then Congress objected and in 1983, Congress denied funding for the MX, the first time a major weapons system request had been refused by Congress. In the following year however the funding was restored.

Congressmen and Senators are usually unwilling to cancel programmes, and if they are doubtful about or hostile towards a Pentagon proposal, the usual response is to vote less funding than is requested. Consequently 'stretchouts' are common, but if the defence contractors and the Services persist, they can usually get major programmes through in the end.

Congressional opinion, in general, has not been hostile to nuclear weapons, although recently, with the growth in concern about 'gold-plated' military budgets, $300 ashtrays, and scandals among the defence contractors, the demand for military value for money has become politically important. Patriotic values and hostility to the Soviet Union are much stronger in the United States than in West European countries, and they are strongly reflected in the Congress. The Chairman of the key Armed Services and Appropriations Committees tend to be conservative, senior figures. For many years Committee Chairman like Senator 'Scoop' Jackson and John Tower used their positions to support the Services' requests, sometimes combining with the Services against the top civilians in the Department of Defense. If the Secretary of Defense decides to reject a particular weapon, for budgetary or other reasons, the Service concerned can make an 'end-run' in the Congress, persuading a Committee to tag on to the budget an amendment voting extra funds for the weapon. The Committee Chairman are not re-elected regularly and many have had long terms of office. Some of them, like Jackson, have acquired great expertise and power in strategic weapons, out-lasting Secretaries of Defense and Presidents. They have real power in the complicated in-fighting which surrounds most major nuclear weapons decisions in the USA.

There have been 'hawks' and 'doves' in the Congress. Senators Brooke and McIntyre fought the Nixon Administration to two successive compromises on the issue of Anti-Ballistic Missiles. Others, like Senator Barry Goldwater, persistently support demands for greater military spending and back programmes like the B-1 bomber. They express their views in a trenchant American style, remote from the verbal deftness and understatement of the British Select Committees. The 1983 hearings of the Procurement and Military Nuclear

Systems Subcommittee of the House of Representatives Committee on Armed Services were begun by the Chairman in the following maner:

Chairman: The subcommittee will come to order.

We start these hearings under a rising crescendo for an instant freeze on nuclear weapons testing, production and stockpiling. We have physicians who tout the horrors of nuclear war and we have other learned groups who have concluded that it would be danger-ous to one's health to be at a nuclear ground zero.

I don't recall their conclusions on the effects on humans who happen to be at a ground zero when a terrorist bomb explodes in a car or an airport locker.

I question such 'spontaneous' activities because the bottom line always is, and I mean always, that it is somehow the fault of the United States, that nuclear weapons exist, that we are the ones who ought to be chastised for their existence.

And the effort that has been directed with such intensity here in the United States is somehow not being directed at all toward the Soviet Union, which has a preponderance of that type of weaponry today.

Although those who support these demonstrations won't agree with it, it is always inferred, if not directly stated, by the leaders of these movements that the United States 'let the nuclear genie out of the bottle', and if the United States would somehow decide unilaterally to give up our nuclear weapons, all would be right again with the world.

That is pure KGB drivel.

An important influence on Congressional behaviour has been the effect of their decisions on defence contractors' plants and hence on employment in their own constituencies. Prime defence contractors deliberately spread the work on major projects to sub-contractors with plants in politically important areas, such as the home areas of members of critical Congressional committees, and then organise lobbying and letter and telephone campaigns to support funding for contracts. Contributions to political committees and other forms of patronage contribute to what has become known as 'the pork barrel' effect. Since Congressmen and Senators are unwilling to be seen to be opposed to national security, and there are always those with strong local reasons for supporting any given weapons project, Congress rarely rejects a weapons system outright. This is particularly so when a programme has reached an advanced stage, when hundreds of millions of dollars rest on the funding decision, and the lobbies in favour of the funds being provided are correspondingly determined. At earlier stages, before a weapon has became a major item, the tendency is for Congress to ignore it in its hearings, since the sheer mass of weapons and line-items in the budget makes it impractical to examine them all.

In summary, Congress does sometimes influence weapons programmes and sometimes slows some of them down. It brings to light a good deal of information. However its record in putting a stop to programmes, even when they are clearly unpalatable to Congress, is chastening. Between 1971 and 1976, Congress cut military budget requests by between 3 and 6 per cent. Of these cuts, 9 per cent were in non-critical areas, 35 per cent were illusory cuts or financial adjustments, 20 per cent were not cancellations but postponents and only 36 per cent (or 2 per cent of the funds requested) constituted real cuts. The great majority of this 2 per cent were attributable to requests for funds for ineffective or duplicate projects.

*   *   *

In representative democracies, parliamentary institutions are the vehicle through which people can influence the government. If nuclear weapons decisions are to be accountable to the public, effective parliamentary control is essential. But at present, in Britain, and still more in France, the role of parliaments is so weak that the decision-making élites can carry out their policies behind closed doors, with no requirement to explain or justify their decisions before they are made. Relative to the US Congress, their powers of scrutiny are limited and their financial control is inadequate. In the Soviet Union and China, where there is no attempt to build a balance of powers into the political system, public accountability on the nuclear issue does not exist. In the United States, in contrast, there is a relatively open and democratic system, with multiple opportunities for Congressional intervention in decisions and a very full system of Congressional financial control. This co-exists with industrial and military lobbies which systematically promote the development of new weapons.

It may be that, in order to restrain the development of nuclear weapons, greater public pressure for restraint, effective political control over nuclear weapons, and an accountable political system are required, as well as international agreement. This combination of conditions exists at present in none of the major nuclear nations.

9 Financial Control

[How financial control over nuclear weapons is exercised: the role of the Treasury in Britain – the budget process in the United States – central planning and financial discipline in the Soviet Union – the importance of financial constraints]

In 1639 King Charles I of England raised an army to fight an invasion from Scotland, and summoned a parliament to collect the money to pay for it. Parliament refused, and nine years later, Charles was executed and the monarchy was overthrown. In the times when there were no standing armies, princes encountered constant difficulties in raising money for defence, and disputes over the control of military spending stimulated the development of parliamentary government.

Three hundred years later, the United States embarked on the Manhattan Project to build the first atomic bomb. The project cost some 2.2 billion dollars and was the most expensive single scientific project ever undertaken to that date. The American taxpayers who financed the project did not even know they were doing so. Subsequently, it has been estimated, the nuclear powers have spent some 3 to 4 trillion dollars ($3–4 000 000 000 000) on nuclear weapons.

The thousands of millions of dollars, pounds, francs, roubles and yuan that are expended every year on the production of nuclear weapons come from revenue derived from taxes on income and production, contributed by ordinary citizens. How do the nuclear nations decide what to spend on nuclear weapons? And how effective are financial constraints in controlling nuclear weapons? These questions are particularly relevant in an age when modern military research and development projects are often subject to escalation in costs. In order to answer them, it is useful to trace how these funds flow from the point of collection of taxes to the facilities which produce the components of nuclear weapons.

*   *   *

In Britain, estimates for the cost of new nuclear weapons are made by the Ministry of Defence. These are incorporated into Long Term Costings, which project forward the total planned expenditure of the

Ministry for the next ten years. The Long Term Costings are the basis of the Ministry of Defence's bid for resources, which is considered by the Treasury and an interdepartmental committee of finance officers called the Public Expenditure Survey Committee. This committee reviews the expenditure plans of all ministries and suggests means of reconciling them. Its report is passed on to a Cabinet committee, which takes the final decision on how much money each ministry will get. The government 'then publishes its expenditure plans in the Annual Estimates. Parliament then approves the estimates. The Treasury, which collects the government's revenue, then makes payments to the spending ministries. Its specific approval is required for payments on nuclear items.

The first stage in the process is the preparation of Long Term Costings by the Ministry of Defence. This is a key planning document, which does not become public. It is prepared by the Office of Management and Budget, the MoD's central finance department. It represents a reconciliation of the financial requests of the three Services, usually achieved not by the imposition of central priorities by the Minister but through bargaining among the Services themselves.

The way in which the Trident project has been financed makes a good illustration of the British process. Trident appeared in the MoD's 1979 Long Term Costings before the government had formally made the decision to go ahead with the project. A central reserve item was set aside as a contingency fund. It was not named, since the Labour Government was still in power, but it was understood in the Ministry that this was its purpose.

The Navy successfully insisted that Trident be kept out of the Navy budget, fearing that its other expenditure would suffer if Trident's costs escalated. Until the 1981 Defence Review, Trident was itemised in the Long Term Costings as a central, national priority item. Thereafter it was put in the Navy budget. Consequently, if the project's costs escalate, the conventional Navy will be likely to feel the pressure first, although if the escalation is sharp, money will presumably be found from central government sources rather than from within the Navy Budget.

When the Long Term Costings have been agreed within the Ministry of Defence, a committee of officials translates the financial requests of all ministries into a document called the Public Expenditure Survey, which forms the basis for the Cabinet's review of planned expenditure. The committee which undertakes this review has become known as the 'Star Chamber'. Its function in every year since 1976 has been to decide on cuts in planned government spending. In defence, however, real spending was allowed to increase by 3 per cent per year until 1985–86, in line with a NATO decision. The Star Chamber resolves the remaining conflicts between ministries' spending plans and the Treasury's limit for total public expenditure, by a process of bargaining between Ministers. The process is intensely political and there is no systematic attempt to weigh against one another by objective criteria all items of government spending. Certainly expenditure on nuclear weapons is not systematically reviewed during this process.

At the final stage of the budgetary process, the Expenditure Estimates are presented to Parliament for approval. The House of Commons debates the Estimates for a few days, but this does not constitute a rigorous examination of the detail of government spending, since detailed information of a kind which would make effective financial control possible is not provided about nuclear weapons programmes. Nor is Parliament presented with information about planned expenditure in following years, so there is no way of assessing what commitments in future years are entailed in the present year's spending.

After Parliament has voted the funds, further formal approval is required from the Treasury before the MoD can spend them on nuclear weapons. In the Treasury, defence spending is dealt with by the Defence Policy Manpower and Material Group, known as DM. DM has a senior staff of 24, who can also call upon an economist, a management accountant and an operational research specialist for help. These officials have the responsibility for monitoring the Defence budget and defence expenditure, running at around £18 billion in 1985–86. DM is divided into two sections, and nuclear weapons come within the responsibility of a subsection within one of them, DM1.

The Treasury's primary interest is to keep overall defence spending within agreed limits. Mr Hansford, a former head of DM, wrote to the Defence Committee on Trident: 'As I am sure you are aware, the Trident programme will be financed within Defence Budget ceilings, which will be controlled as part of total public spending under the terms of the Government's medium-term financial strategy. . . . Expenditure on Trident will be contained within the Defence allocations.' The nuclear force represents a part, and a rather small part, of the entire defence budget. In practice, the Ministry of Defence has considerable autonomy over how it allocates its funds internally. For example, the Navy took advantage of an underspend in the Trident budget to use Trident moneys for other purposes. It seems possible that the same could be done in reverse.

The control that the Treasury can exercise is limited by the priority given to nuclear strategic projects by the Government. This leads the Treasury to take the view that, if the Ministry of Defence requires a particular system, the money has to be found. Hence the sharp rise in Trident's estimated total costs in 1984, from £5 billion in 1980 to £10.7 billion in 1984, appeared to cause little anxiety in the Treasury, since it related to future expenditure which, it was assumed, would be contained within planned totals by decisions to be taken in the future. In the meantime, expenditure on the early contracts and on the development of warheads was approved.

In 1985, the government decided not to continue increasing defence expenditure at 3 per cent per annum, but instead to hold it about level in real terms. This decision will increase the financial pressure on the Ministry of Defence, which is now attempting to maintain four defence commitments, all subject to rising costs, within a fixed budget. These are the British Army on the Rhine, the defence of the British Isles and the Atlantic Approaches, the independent nuclear force, and commitments outside NATO, including the Falklands. As the Trident programme enters a phase of increasing costs, pressure on conventional equipment procurement is bound to be felt, and this will intensify the struggle for resources within the Ministry of Defence.

Financial pressure of this kind can, potentially, be an important constraint on nuclear weapons, through overall defence spending limits. But senior civil servants do not expect the Treasury to succeed in controlling individual projects. 'One Treasury insider reckons that the DM team are well-intentioned, highly intelligent, they huff and puff, but there is a defeatist streak. Their job is to put up a good fight but not in the end to win . . . the MoD has three hierarchies of civil servants, military and scientists, all difficult to penetrate and all skilled at special pleading.'

Ultimately, major spending decisions, which affect whether a particular system should go ahead, are taken by a Cabinet subcommittee such as MISC7. For example, the decision to proceed with Trident as it became more expensive was taken by Ministers. The decision to keep Chevaline going was taken at a sequence of meetings of a Cabinet committee which agreed to several successive tranches of short-term funding. There have also been occasions on which Ministers have cut nuclear projects on financial grounds. In 1960 the British-designed intermediate-range missile 'Blue Streak' was cancelled when its costs began rising sharply, though obsolescence was also a factor in that decision. In 1965 the Wilson government cancelled the nuclear capable aircraft 'TSR2', in order to implement cuts in defence spending. These Cabinet decisions are taken at a broad strategic level, and Ministers do not always have the information, the technical knowledge or the time to monitor projects in detail. Expenditure on Chevaline was approved for several years with little grasp among Ministers of what the project really involved. Detailed financial control was delegated to the official in charge of acquiring the weapon. The official in charge of the Trident project, the CSSE (Controller of Strategic Systems Executive) is also charged with financial monitoring and reporting of the Trident programme, assisted by a Trident Finance Office. The reports go to the Trident Watch Committee which consists of senior decision-makers responsible for advancing the project.

Finally, after the expenditure has been made, the most rigorous arm of Parliamentary control comes into play through the auditing of public accounts. This is the responsibility of the Comptroller-General, who reports to the Public Accounts Committee. In the case of Chevaline, although the Comptroller-General was aware of the project and its escalating cost, he did not inform the Public Accounts Committee of this. The Public Accounts Committee has now arranged to see details of major spending on the Trident project.

In principle, all expenditure on nuclear weapons produced in Britain is voted through by Parliament and approved by the Cabinet, with Treasury control of specific expenditure and further financial control in the Ministry of Defence. In practice, given Parliament's lack of information, the exclusion from the process of most of the Cabinet, the deference of the Treasury to the Ministry of Defence, and the internal struggle within the Ministry of Defence which deprives its financial controllers of final authority, the effective financial control of nuclear weapons is shared between a small inner circle of Cabinet Ministers and those civil servants, nuclear scientists and military officers within the MoD who are responsible for their acquisition. There is little financial control of nuclear weapons independent of those decision-makers who are responsible for advancing nuclear projects. The limited role of Parliament in the system of financial control is especially striking, particularly in comparison with that of the Congress in the US system.

*   *   *

In the USA every major nuclear weapons programme has to secure approval every year from the military Service which is developing it, the Secretary of Defense, the Office of Management and Budget in the Executive, and Congress. For delivery systems, there are two main forms of financial control: through the budget, which supplies the Services with the funding needed for weapons programmes, and through the acquisition process, which controls the placing of major contracts.

The budgetary process starts 21 months prior to the beginning of each financial year, and involves a formidably complicated process of review by different government bodies. It starts when the Secretary of Defense issues the 'Defense Guidance', which outlines the general principles to guide the military budget. For example, the document for the fiscal year 1984 set out the following principles: the US Services should be prepared to fight a nuclear conflict that might last several months; the Services should pose a credible conventional threat to Soviet forces in many geographical regions simultaneously; the Navy and Air Force should develop co-operative defensive measures to protect naval fleets from Soviet air attack; military services should put greater emphasis on special operations such as guerrilla warfare, sabotage and psychological warfare.

The Joint Chiefs of Staff then set out the military requirements for new weapons, in the Joint Strategic Planning Document. At the same time, the Commanders-in-Chief of the eight US unified and specified commands offer their views on threats, military strategy and budget needs. These statements of military requirements have an important influence in shaping the character of the entire budget process which follows.

Next, each military Service draws up its budget request in the form of a five-year 'program' specifying the weapons to be procured, forces to be deployed and expenditure to be incurred in the next five years. It is here that requests for new weapons systems enter the budgetary pipeline.

The Secretary of Defense and his civilian managers then review all the Service requests and draw up the Department of Defense Budget Request. He is aided by a high-level body called the Defense Resources Board which resolves many of the budgetary conflicts between the Service requests. Key individuals within the Defense Resources Board are the Under Secretary for Policy (currently Richard Perle), the Chairman of the Joint Chiefs of Staff, and the Under Secretary for Research and Engineering, the official responsible for developing nuclear strategic systems.

The Secretary's Budget Request is then passed to the Office of Management and Budget, which reconciles the military budget with domestic spending plans on behalf of the President. Because the military budget is so huge, a practice has grown up that staff of the Office of Management and Budget are permanently seconded to the Pentagon. The OMB only has about 27 examiners to review the defence budget, so each of them has over $10 billion of military programmes to review. A clash between OMB director David Stockman and Secretary of Defense Caspar Weinberger during the Reagan Administration led to Stockman's resignation and a reduction in the influence of the OMB over defence spending.

The President and his advisers can, if they wish, make further changes, and occasionally the National Security Council may consider aspects of the defence budget on an *ad hoc* basis. When the President has approved the budget it then passes to Congress.

In Congress, the first stage is consideration by the Congressional Budget committees, which place ceilings on the global budget and a subceiling on 20 spending categories known as 'functions'. Function 050 is 'National Security' and includes all the Department of Defense military programmes and the Department of Energy's nuclear warhead programmes. The Senate and House Committees on Armed Services and Appropriations proceed to review the programmes in the military budget and the appropriations required. Towards the end of the process, conference are held between these committees to reconcile conflicts; and sometimes the President will meet Congressional leaders to seek an acceptable amended budget. Frequently during this process, items which have been cut by one committee or another are restored. Even if Congress turns down a request for funding a programme, the President can ask for a supplemental appropriation for the year underway to restore them. In January 1983 President Reagan requested additional funds for production of the MX in fiscal year 1983, which had been refused by Congress the year before.

Despite this lengthy process of scrutiny and oversight, in practice the wishes of the military services are normally little changed by Congress and the Executive. In the 1960s only about 4 per cent of the programmes in the military budget were altered by Congress, and about 19 per cent changed by the Executive. The fact that the Services and the defence contractors can lobby for a weapon system as it passes through each stage of the budgetary process weakens the independence of successive reviews. Pressure continues to be applied even in the Office of Management and Budget, the President's budgetary review body. For example in 1983 the B-1 nuclear bomber was in danger of losing the Armed Services Committee vote on multi-year procurement because of the rocketing cost projections of the programme. Rather than risk the programme being lost, Republican members on the committee persuaded the Office of Management and Budget to adjust upwards the projected figures for income available for defence spending, by an amount which matched almost to the dollar the figure needed to secure the B-1.

The Reagan Administration's large increase in defence spending may be curtailed after 1986 by a Congressional measure aimed at balancing the Federal budget. The Gramm-Rudman amendment, which has been signed into law by President Reagan, requires the Administration to balance the budget by 1991. Unless the courts declare that the law is unconstitutional, if the government fails to balance the budget itself, the Gramm-Rudman law could require across-the-board cuts in defence programmes, which could restrict production of Trident submarines, MX missiles, B-1 bombers and cruise missiles. If Gramm-Rudman has this effect, it would unintentionally achieve more restraint on US nuclear weapons than either arms controllers or opponents of nuclear weapons have been able to achieve to date.

Even if the Gramm-Rudman law takes effect on the Pentagon's programmes, it is likely that the design of new nuclear warheads would escape its impact, since this is financed by the Department of Energy, not the Department of Defense. Financial control of warhead production is particularly weak because of the split in responsibilities between the two concerned Departments. The Department of Defense places the orders for the number and type of warheads it wants, while the Department of Energy has the responsibility for managing the nuclear weapon production facilities and paying for them. Consequently, as Senator Sam Nunn said in the Senate Armed Services Committee, 'there is a built-in incentive for DOE to build the most expensive warhead possible and to build as many as possible. The Department of Defense is not constrained to consider cost in setting warhead requirements because DOE funds the warhead cost'.

In addition to the budget process, there is a separate process for the control of acquisition, administered within the Department of Defense. This is the DSARC process which reviews major programmes costing over $200 million in research and development or $1 billion in production costs. Lesser sums are largely under the financial control of the Services themselves, which have considerable autonomy in the management of their own budgets – a practice which has been encouraged by recent Administrations. It is relatively easy for the Services to move funds from one budget line to another and thus continue cancelled projects, start new projects using remaining funds from old ones, and so on.

Practices such as these enable the military Services to evade Congressional financial control. In 1983 for example Congress deleted funding for the Pershing II missile. But the Army kept the funding flowing anyway, taking money out of other missile programmes to sustain the Pershing production line. The Pershing has been one of the missiles with the greatest impact on East-West tensions in recent years because of its ability to strike in only eight minutes at protected targets – for example it could destroy command and control centres in Eastern Europe and the western part of the USSR in a sudden 'decapitating' attack. The proposal for the weapon originated in an unsolicited proposal by the prime contractor, Martin Marietta, financed by an 'Independent Research and Development' subsidy from the Pentagon which is used to fund retrospectively independent research by the contractors. Congress has no control over these funds.

* * *

In 1983, two private US think-tanks, the Hoover Institute and the Heritage Foundation, advised the Reagan Administration that the US could 'win the arms race' if it forced the Soviet Union to spend so much on armaments that its economy would be bankrupted.

It is not reported whether these two institutes considered the likely response of the Soviet Union to such an attempt. The Politburo has been absolutely determined, since Stalin's time, to acquire a sufficient military force to resist attack from the West and, since Brezhnev's time, to maintain 'rough parity' in strategic arms. It is difficult to imagine circumstances in which the Soviet Union would fail to match US strategic initiatives, irrespective of the cost. However, one possible difference between the two countries is that the Soviet Union might be more likely to know when it was being bankrupted than the USA. This is because of fundamental differences in their systems of economic and financial planning.

In the Soviet Union all investment is carried out by the government, under the guidance of five-year plans formulated by the central planning agency, GOSPLAN. This body considers the material needs of all the major sectors of the economy, including the defence sector. Working under the general guidance of the Politburo, GOSPLAN stipulates planned levels of output for the defence industries and for all the domestic economic sectors. As a result, the Soviet government is acutely aware of how its investments in the military sector are at the expense of investments in civil, social and economic programmes. In the West, where economic planning is more diffuse and different interests have economic stakes in military and civil programmes, the choices do not appear so clear-cut.

Khrushchev clearly saw this conflict when he said, 'I decided that we had to economize drastically in the building of homes, in the construction of communal services, and even in the development of agriculture in order to build up our defences. I went so far as to suspend the construction of subways in Kiev, Baku and Tbilisi so that we could redirect those funds into strengthening our defence and counter-attack forces.' In 1985 the new five-year plan once again emphasised heavy industry, engineering and machine tools. The five-year plan requires Soviet leaders to make choices between investment in defence-related industries and the rate of growth of personal incomes and consumer spending. Economic choices of these kinds are rarely made explicitly in the West.

The Soviet military, of course, press for new weapons in the same way that the US military does. Khrushchev refers to this in his diaries: 'I'm not saying there's any comparison between our military in the socialist countries and capitalist generals, but soldiers will be soldiers. They always want a bigger and stronger army. They always insist on having the very latest weapons and on attaining quantitative as well as qualitative superiority over the enemy. . . . That's why I think that the military can't be reminded too often that it is the government that must allocate the funds; it is the government that must decide how much the Armed Forces can spend . . .'.

In the Soviet Union the government has much more control over how the Armed Forces spend the funds, since the government not only sets the military budget but also determines the prices of military equipment and the output levels of the industries which produce it. The centralised process of decision-making over weapons acquisition makes it virtually impossible for the military to exercise autonomy in its weapons-buying policies or to ignore and subvert the decisions of the political authorities.

The Politburo has the ultimate responsibility for approving the budgets and deciding on the level of forces and the number of weapons. These are based on detailed budgets and plans drawn up by the Defence Ministry. The General Staff has a central financial directorate which has an important role in resolving the wishes of the separate military Services within the overall budget limit. There are also inevitable adjustments during the course of a five-year plan, which require changes in the allocation of scarce materials and industrial resources. These are resolved in part by the General Staff, which may reorganise priorities among the military Services, and in part by the Military Industrial Commission, a top-level political body which supervises the defence industries as a whole. When weapons programmes experience major cost overruns, the Politburo reconsiders the project, and nuclear weapons systems have been cancelled because they have exceeded their planned costs.

The Soviet system operates as a single large corporate body, taking a relatively long-term approach to its management of nuclear weapons, with internal planning of both finance and material resources. The contrast with the United States, with its annual budgetary battles and its semi-autonomous military and government bodies and autonomous corporations, is striking.

*　　　*　　　*

Financial constraints have been the most important restraint to date on the development of nuclear weapons – certainly more important than public opinion, parliamentary opposition, or arms control.

Yet financial control of nuclear weapons is weak. In the Western countries it is exercised by central financial institutions which lack sufficient staff and sufficient information to subject highly complex and technical programmes to searching financial review.

The British Parliament exercises little real financial control and the French Parlement still less. Congress, despite its elaborate and detailed procedure for review of the budget, rarely alters the requests of the Services, and when it does, its decisions are frequently overturned in subsequent years. Consequently in Britain, France and the USA, effective financial control of nuclear weapons has sometimes been delegated to the very departments which are responsible for their acquisition.

In the Soviet Union too, detailed procurement matters are largely in the hands of the professional military. In the Soviet Union however, the Politburo and its ancillary institutions exercise real control over the military budget as a whole and discontinue individual weapons programmes in which cost escalation is severe.

In all five nuclear states, the claims of finance are forced on the weapons-makers to secure value for money and cost-effective weapons. Financial restraints do limit what systems these states can afford to deploy. However, the development of nuclear weapons is considered such an overriding priority that the money has to be found. Often, therefore, it is the requirement for new nuclear weapons which sets the level of the financial constraints – not the financial constraints which set the level of nuclear weapons.

10 Public Opinion

[The big demonstrations – public opinion and its impact on nuclear weapons decisions in Britain – the lack of debate in France – the grass-roots and the lobbyists in the United States – patriotism and dissent in the Soviet Union – the first anti-nuclear demonstration in China – the role public opinion could play]

The centre of the city was jammed with people. Snaking in a long line through the streets, crossing several bridges over the river in different directions, the procession stretched for over a mile. Carrying placards and banners, shouting, singing, banging drums, the demonstrators marched along. Half a million people had come and they had brought the centre of the city to a standstill.

This scene, duplicated in London, Amsterdam and New York, marked the depth of concern about nuclear weapons policies in the Western countries. There is no doubt that the deployments of Cruise, Pershing and SS20 missiles and of other new weapons has triggered a wave of public alarm, especially as the climate of international relations continues to deteriorate. What has been the effect of public opinion and debate on those who take decisions about nuclear weapons?

*　　*　　*

Of all the five major nuclear weapons states, Britain is the one which has had the most long-standing and intense debate over nuclear weapons. The Campaign for Nuclear Disarmament (CND) was the focus of a mass movement of opposition to government nuclear policies in the 1950s and 1960s, and again became a powerful force in the late 1970s and 1980s. Britain is the only country of the five in which a major political party has declared that it will get rid of nuclear weapons if elected to power.

Public opinion is clearly important, since elections are conducted on the basis of polls of the whole adult population. It is difficult to establish, however, what public opinion is. There are about 44 million adults in Britain, who hold a wide range of opinions. Some hold their opinions with greater intensity than others and some are better informed than others. Professional analysts of public opinion distinguish between the mass public, the attentive public, the opinion leaders and the élites. It is the mass public whose opinions are sought in public opinion polls, but this includes many people who may not have voiced their opinions on the subject in question and may not be concerned about it. Nevertheless, this public is important as a test of electoral opinion. The attentive public are those who take an interest in the subject and are well informed about it. Opinion leaders are those who form opinions among either the mass or the attentive public. The élites are those who hold top positions in politics, business, the Civil Service and other spheres. Elites may hold opinions different from the rest of society but are highly important, since those who take decisions on nuclear weapons are themselves an élite, recruited from an élite and more influenced by élite opinions than others.

Over the past two decades, opinion polls have shown that about two-thirds of the population believes that Britain should retain nuclear weapons. In January 1985 for example, a poll showed that 65 per cent thought it would be a mistake for Britain to give up relying on nuclear weapons for defence.

The polls also suggest that the public dislikes the acquisition of new nuclear weapons. Polls have shown majorities against the deployment of cruise missiles and against Trident. A poll in November 1985 indicated that almost three-quarters of the public would favour a world-wide freeze on new nuclear weapons. Moreover, 84 per cent thought that, if the Soviet Union were to stop nuclear testing (as it did in 1985), Britain and the USA should reciprocate.

Pessimism about the likely effectiveness of nuclear deterrence is widespread, with polls in 1980 and 1983 suggesting that 39 per cent and 49 per cent respectively believed a nuclear war was likely to occur. Support for neutralism is surprisingly high. For many years the polls have indicated that about 50 per cent have little or no confidence in the United States' ability to deal wisely with world problems, and about 45 per cent think that Britain should be neutral. 50 per cent would prefer to see Britain behaving like Sweden or Switzerland rather than playing the role of a world power.

These figures are interesting, but it is questionable how much opinions of this kind influence political change. At General Elections, voters choose between parties on the basis of overall policies and party images. For most voters, defence is an issue of lower priority than economic matters. Although defence became more salient in the 1980s and the retention of nuclear weapons became an election issue in 1983, it is impossible to distinguish the importance of the public's views on nuclear defence issues from the many other issues raised.

Through elections, the public determines which party is in power, and the leader of the party which commands a majority in the House of Commons becomes Prime Minister. The Prime Minister then selects the Minister of Defence and the other senior Ministers who will become the only politically accountable persons involved with decisions about nuclear weapons. Given the way the electoral system operates in Britain, the outcome of elections is determined by the swing of opinion among undecided and uncommitted voters, and it is to this group that the rival parties appeal. General political moods, feelings of economic insecurity and attitudes towards the party leaders probably have a greater impact on determining which politicians run defence policy than public views on defence.

For setting the positions political parties adopt, specific constituencies of 'attentive' opinion are more important than mass public opinion. The Labour Party is influenced by opinion in the Trades Unions and its constituency parties, and it was these groups, more radical than the electoral support for the Labour Party, which secured the Labour Party's commitment to unilateralism in 1982. The Conservative Party has a strong constituency in the Armed Forces and among businessmen. A sharp antagonism towards communism and a traditional hankering after British greatness make it a party naturally disposed to favour strengthened nuclear forces.

Other than at elections, public opinion has, hitherto, had little discernible effect in Britain on nuclear decisions. A majority opposes Trident, but the government ignores it. The only majority that counts is in Parliament.

If the effect of public opinion on the top politicians is indirect, its effect on nuclear decision-makers in the Ministry of Defence and Aldermaston and the defence industry is negligible. The Ministry of Defence is an extremely closed body. Even respected think-tanks like the International Institute of Strategic Studies, the Royal Institute for International Affairs and the Royal United Services Institute have little impact on policy, since they have no access to classified information and no prior knowledge of governmental policy deliberations on nuclear weapons decisions. General public opinion makes still less impression.

*　　*　　*

In April 1985 a gigantic figure on stilts passed through the streets of Paris. Dressed as a bureaucrat and carrying a briefcase, he was driving three smaller figures on leads – a white-coated scientist, a uniformed military officer and a pinstripe-suited deputy of the Assemblée Nationale.

This figure was the centrepiece of a demonstration against French nuclear weapons which had for its motif the question: 'Who decides for us?' Certainly, whatever answer to the question is correct, public opinion in France would not be a part of it.

One of the most striking characteristics of the French political environment has been the absence of a public debate. Opposition to nuclear weapons has been dominated by the French Communist parties, which have been compromised in the eyes of most French people by their sympathies with the Soviet Union. Other than the Communists, the major political parties all approve of French nuclear weapons. While there is disagreement on points of detail, there is a solid consensus behind the government's basic nuclear defence policy. The Catholic and Protestant Churches also support this consensus.

Perhaps because of the relative lack of political debate, the public mood is one of disinterest. Opinion polls suggest that only some 9 per cent of French people feel personally concerned about issues of international peace and war. The polls suggest a strong vein of cynicism: while 52 per cent believe that the French nuclear force is necessary for France, 40 per cent believe it is a waste of money and over 58 per cent would oppose the actual use of nuclear weapons, even if France were invaded. At the same time, 44 per cent believe that disarmament will never be possible. The polls also suggest that there is strong support for conventional national defence and a fear of nuclear weapons.

*   *   *

In the United States nuclear weapons were a subject of intense debate from the very earliest days of their development, even before they became publicly known. The controversy began among the scientists at Los Alamos over whether the bomb should be used in the war. It grew after the war as a group of scientists sought to place nuclear weapons under international control. Later there were heated debates in the scientific community over whether or not the USA should develop the H-bomb. Nuclear weapons remain a subject of intense debate among the scientific community in the USA, in striking contrast to the situation in the United Kingdom where the scientific community eschews political controversy.

The United States of America is often called a 'pluralist' society. With its big cities, the vast prairies, the humid southeast and the desert southwest, the American people are a mix of backgrounds and ethnic origins. Opinions on nuclear weapons range widely for those of the religious fundamentalists and extreme groups on the right such as the Survivalists, to the anti-war groups which became an effective coalition during the Vietnam War. Despite these varieties, Americans have a strong sense of American identity and patriotism and these values have been linked by political leaders of both American parties to the maintenance of military strength together with the defence of American and Allied interests throughout the 'free world'.

American opinion polls show that Americans have a strong sense of being threatened by the Soviet Union, stronger than people who live much nearer the Soviet Union in Western Europe. A poll in 1982 indicated that 84 per cent of Americans see the Soviet Union as a 'threat to the security of the United States'. Sixty-three per cent thought that the Soviet Union would attack the United States some time in the next forty years. Forty-nine per cent viewed the Soviet Union as an 'outright enemy'. In 1980, 83 per cent considered war likely.

About 63 per cent of Americans favour introducing 'just enough nuclear weapons to create a balance between East and West until an acceptable agreement can be found', according to a poll in October 1983. The same poll showed that about 20 per cent of Americans favoured introducing 'more nuclear weapons than the Soviet Union has introduced, in order to establish and maintain nuclear superiority'. Eight per cent thought that the USA should 'introduce no more nuclear weapons, even if the Soviet Union does', and 4 per cent were in favour of giving up nuclear weapons unconditionally.

Support for a multilateral freeze on the production of nuclear weapons was very high in the 1980s, with over 70 per cent subscribing to the idea. Americans also favoured persisting with the Strategic Arms Reduction Talks (START) by a wide margin. However, the polls suggested that 78 per cent believed that 'the Soviets only want agreements when they can gain an advantage', and 75 per cent were sceptical that the Soviets would keep to the agreements.

These figures help to explain why promises to 'keep America strong' are made by both Democrats and Republicans in American elections. As in Britain, it is difficult to distinguish the significance the electorate attaches to this issue compared with others. Domestic and economic issues influence voters more than matters of foreign policy and defence. Nevertheless 'missile gaps' and scares about American strategic inferiority have featured in several elections, to the benefit of the successful challenger in the cases of Kennedy, Nixon and Reagan.

In a culture steeped in advertising and the 'hard sell', it is perhaps not surprising that defence contractors and other private interests promote their products vigorously and openly manipulate public opinion. Company presidents lobby their employees and local residents to support their programmes. One such appeal was mounted by Rockwell: 'If you agree with the experts, write, wire or telephone your Congressman today and urge him to support national defense and the B-1 bomber which the experts say is needed for the security of our country. Also, I hope you will urge your family, neighbours and friends to take similar action.' Pressure groups in favour of greater defence spending organise sophisticated lobbying campaigns which include the sending of personalised mail-shots from retired generals to selected Congressman and opinion-leaders, telephone campaigns, briefings, and radio and TV appearances. Similar campaigns are also mounted by groups opposed to nuclear weapons or increased defence spending, but the lobbies in favour command and expend much greater financial resources.

On two occasions public pressure has played a role in stopping nuclear weapons systems. The first was in the controversy over the deployment of an Anti-Ballistic Missile system, when local opposition in the area where the missile system was to be sited had a powerful influence. The second was over the Air Force plans to base the MX missile in a large number of shelters with connecting roads scattered over seven thousand square miles in Nevada. Opposition from ranchers and Republican local residents forced the government to abandon the scheme. Public opinion was also influential in the adoption of the Partial Test Ban Treaty, which was seen to limit the dangers to health from fall-out from atmospheric tests.

It has been when the personal interests of Americans have been directly affected that public opinion on nuclear weapons has made an effective impact. On more generalised issues public opinion has failed to have an effective impact. For example, the widespread grass-roots support for a nuclear freeze won considerable support in the State Legislatures but then became blocked and ran out of steam in Congress.

Major new weapons programmes, such as the MX missile and the Strategic Defense Initiative, have become the subject of public controversy and debate. In turn, the public debate and comment in the media has an effect on opinion in Congress. However, the more influential debate is not the one in public, but the one that goes on in the weapons community of defence scientists, defence contractors, government officials and military officers. Much of the information used in these circles is classified and assumptions are shared which do not reflect public opinion as gauged in the polls. It is the opinion of this community, circulated through strategic think-tanks, advisory committees and defence science boards, which influences the policy the government actually adopts. Over most nuclear weapons, if public opinion becomes engaged at all, it becomes aroused only when the programme is sufficiently far along its development path to have acquired a lobby in its favour and a rationale that is ready for 'selling' to the political system. The Strategic Defense Initiative is a striking exception to this rule, but it was introduced in a completely exceptional way, bypassing the weapons community and the Pentagon's usual channels.

If public opinion plays a larger role in the United States than in the other major nuclear states, it is because the role of Congress is more important and the American decision-making system is less secretive. Even so, American public opinion normally has a marginal or negligible impact on nuclear weapons decisions. Moreover, of all the five states, public opinion in America is probably the most susceptible to hostile perceptions of the enemy and a sense of vulnerability to threat.

In an interview conducted in Moscow in 1984, a former Soviet colonel and adviser on international and defence affairs was asked: 'Can there really be a risk of war between the Soviet Union and the West? Is the conflict of interest between the two sides so very sharp?' He replied, 'No, the conflict of interest is not so sharp. There are no significant territorial disputes. Neither side threatens the vital interests of the other. And yet, strange as it may seem, there really is a risk of war. For us in this country, war no longer appears just as an abstract possibility, as it did even a year ago. There is a feeling that things really might come to that. . . . It is the combination of our mistakes and your hysteria that is so very dangerous.'

* * *

In Moscow it is the custom for newly-weds to lay flowers on the Tomb of the Unknown Soldier on the day of their wedding. Elsewhere in

Russia, newly-weds pose for photographs in the town square in front of rocket-launchers dating from the Second World War. Memories of the War are still intense and they are kept alive both by people and the government. The trauma of those years, in which 20 million Russians lost their lives, created a bond between the Communist state and the mass emotions of the people which is perhaps the Soviet government's greatest claim to the people's loyalty.

Military service is compulsory in the Soviet Union. It is preceded by a period of pre-induction training in schools. The Armed Services organise 'military-patriotic education' as well as technical training in schools and colleges.

Military-patriotic education', which is provided to all, is intended to 'instil a readiness to perform military duty, responsibility for strengthening the defence capability of the country, respect for the Soviet Armed Forces, pride in the Motherland and the ambition to preserve and increase the heroic traditions of the Soviet people.' It includes basic military training, learning of technical skills, political indoctrination and the 'development of qualities of will, courage, hardiness, strength, speed of reaction, etc.'.

According to unofficial reports, military–patriotic education is not particularly popular, nor is conscription. Nevertheless, there is little questioning or criticism of military policy in the Soviet Union. The Armed Forces are held in more respect than in Western societies. The maintenance of powerful military forces has a strong appeal to Russians, who remember the successive invasions from the West of their country in the First World War, the Russian Civil War and the Second World War.

A Russian journalist said:

You must try to understand our psychology, not only Soviet psychology but also the Russian psychology which goes far back before 1917. Our extreme defensiveness has always stemmed from an acute consciousness of external danger. Whenever we feel at all under attack, we close our ranks and make sure we present a united front. And we take talk of the 'evil empire' much more seriously than you do, or than you expect us to.

Our society is different from yours. Even before the Revolution, our emphasis was on the collective, yours on the individual. You must learn to be tolerant of this difference, not to make our adoption of your ways a precondition of regarding us as a normal country . . . So the impression of a consensus in Soviet opinion, while an oversimplification, is far from totally false. Almost all Soviet people do support the peace policy of our government. The unity of views in this area is much greater than when it comes to our internal problems.

However, there is little opportunity for dissent from official policies to be expressed, especially on an issue so central to Soviet security as nuclear weapons. The Russians are naturally guarded in their expression of opinions and the risk of informers is very real. It is difficult to judge public opinion when it cannot be expressed.

The first sign of a challenge to Soviet nuclear weapons policy came in 1982, when an independent Group to Establish Trust Between the USSR and the USA was set up in Moscow. The group held a number of candle-lit vigils and called for Moscow to be made a nuclear-free zone. The KGB responded by deporting some members of the group, putting others into psychiatric hospitals, and driving some to isolated spots where they have been beaten up.

In this way the Soviet authorities indicate the boundaries of legitimate public expression.

*   *   *

Members of the Turkic speaking Uighir minority make up nearly half the 13 million inhabitants of Xinjiang, China's vast and remote western province which covers a sixth of the country's land area.

Since 1976 there have been reports in the provincial capital, Urumqi, of Uighir men and women dying prematurely at the age of 40 and 50. Children have been born deformed, lambs and calves born without limbs, and grass and crops have been poisoned. The affected areas lie in southern Xinjiang, where 80 per cent of the Uighirs live. Although China's Lop Nor test site is in the north-eastern part of the Gobi Desert, the area is known for its violent sandstorms and high winds.

China has carried out six nuclear tests in the atmosphere since 1976, the last in October 1983. Chinese officials have admitted that they underestimated how far the radioactive dust would carry and failed to move people sufficiently far from the fall-out area. Some victims are receiving pensions of around 10 yuan a month (£3.80).

In 1985 students from Xinjiang held a protest rally in the Tian An Men Square in Beijing. They presented a list of demands to the government, headed with a demand for an end to nuclear testing. This was the first anti-nuclear protest in Beijing and followed several larger demonstrations in Urumqi in which several thousand people participated.

Following the demonstration, the Chinese Foreign Ministry issued a statement saying that about 200 Uighir students had 'put a few questions chiefly because they did not know much about the situation'. 'In the present international situation', the statement went on, 'it is necessary to conduct a small number of nuclear tests to safeguard China's security. This is endorsed and supported by the great masses of the Chinese people.'

There is no anti-nuclear lobby in China, but officials have confirmed that some Chinese feel uncomfortable about nuclear developments and 'an education campaign has been necessary to put their fears at rest'.

The Uighirs have a long history of armed opposition to Chinese rule and their latest Muslim leader was captured and executed in 1961. The students' demands included an end to the Chinese policy of sending criminals to Xinjiang as punishment, democratic elections for minority officials and a relaxation of the one-child birth control rules. One Uighir said: 'There can be no uprising here, but we will continue to protest and make our voice heard to win some real autonomy and one day our own independent nation.'

The Uighir demonstrations are a significant milestone in China, but Xinjiang is a long way from Beijing. At the centre of political power, those who were carrying out China's nuclear weapons policies have little fear of intervention by public opinion. Indeed, despite Chairman Mao Zedong's famous invocation to his cadres to 'learn from the people', China's leaders generally have their own interpretation of what the people want and what is best for them.

The Chinese are traditionally a disciplined people, who accept the need for a coherent and unified line. Once a position has been adopted, the Chinese will follow it together. It is not Chinese, nor is it considered politically correct, for rival schools of thought to openly advocate different courses. Although there are debates within the Party and in the country on issues, once a policy has been decided by the Party, it becomes the correct line. Part of the role of the Party has been to educate the Chinese to understand the correctness of the government's course. Party leaders sometimes encourage the expression of public opinion, for example through the Big Character posters which became popular in the Cultural Revolution, and through local meetings held to discuss national issues. However, when opinions turns against national policy, there is usually a clamp-down. It is mainly when the Party is divided against itself that the expression of public opinion is encouraged.

There are signs that political discipline is becoming a little more relaxed under the influence of Deng Xiaoping's modernisation programme. Freedom of expression and a certain amount of liberalisation are seen as necessary ingredients of modernisation. The Chinese are encouraged to 'liberate their thinking', that is, to hold their own opinions rather than others', and express opinions in letters to newspapers. The attitude is summed up in the view stated in a newspaper leader that 'people should be allowed to express their views, even when they are wrong'.

Little information about Chinese weapons development programmes is published. More emphasis is placed on the large arsenals and aggressive intentions of the superpowers. One well-informed Chinese, for example, said that China has only a few rockets, which are mainly experimental. The danger to world peace, he said, lies in the enormous build-up of weapons by the United States and the Soviet Union.

The National People's Congress has been permitted to debate certain issues, and there has been some public discussion of the hazards of civil nuclear power. But nuclear weapons policy is not a subject on which debate is encouraged, nor does it appear to be, outside Xinjiang, a matter of public controversy.

*   *   *

If nuclear weapons are to be restrained, then public opinion in the nuclear weapons states is one of the most important potential restraining forces. This chapter has indicated that, hitherto, public opinion has had little impact on nuclear weapons decision-making. In the Soviet Union and China, open debate on national nuclear weapons policies has not been permitted. In France, a debate has not occurred. In Britain the debate has been vigorous, but those who have been involved with taking decisions on nuclear matters have not been greatly influenced by public opinion. This has been due partly to the secretive and closed character of the main defence institutions, and partly to the consensus in favour of nuclear weapons which existed, up to 1979, between the political leadership of the two main parties. Only in the United States is there evidence that public debate has actually had an influence on nuclear weapons decisions, and the level of public debate there is more open and better informed than elsewhere.

In the West, public opinion is divided on these issues. A majority appears to favour the maintenance of nuclear weapons, but the nuclear weapons policies currently being pursued by the Western governments do not correspond with public views, as these are measured in the polls. A significant change in public opinion could bring about changes in policy. Despite the widespread sense of powerlessness, which may be influenced by the way in which nuclear weapons decisions are taken, the people have the power to change their governments and their governments' policies, if they choose to exercise it. An aroused public opinion in the nuclear weapons states could be the most important of checks to the development of nuclear weapons.

11 The Arms Controllers

[The walk in the woods – the Geneva summit – a housewife intervenes – the arms control talks – how arms control policy is formulated in the United States – in the Soviet Union – in Britain – in France – and in China – a basis for progress]

It was summer, but overcast. The pure air of the Jura mountains in Switzerland was turning sticky. In the forest, all was quiet. It was beginning to rain. On the forest trail there was a log, and sitting on the log were two well dressed men, one elderly, one middle-aged. Shielding themselves from the drizzle, they bent over the papers on their knees. They were both in a state of considerable excitement. Pencils flew across the papers as they deleted sentences and added clauses. They talked together rapidly, in very different accents. Finally they got up, satisfied. They had reached an agreement.

The two men were Paul Nitze, chief of the US delegation to the Intermediate Nuclear Force talks, and Yuli Kvitsinsky, chief of the Soviet delegation. The agreement they had made would limit the numbers of intermediate-range nuclear weapons in Europe.

The two men were under no illusion about their chances of getting the agreement accepted. They had both gone beyond the terms their governments had set. 'I'll tell them it's your scheme, and you tell them it's mine', said one. 'Maybe we'll both go to jail', said the other.

Within a month, both their governments had repudiated the agreement.

* * *

It was winter, and snowy. The mountains were heavy with snow, and at the lake shore, a thick white layer covered the roofs of the buildings. Inside one cabin, two men sat at a log fire. One was the President of the United States of America, the other General Secretary of the Communist Party of the Soviet Union.

'I simply cannot condone the notion that human beings can only keep the peace by threatening to blow each other away', said the President. He was conscious of Soviet fears that his planned Strategic Defense Initiative was a cover for developing a first-strike nuclear capability, but he was prepared to share the technology with the Soviet Union. The General Secretary looked at him silently for a while. Then he replied. 'I understand what you are saying, but I disagree with it. It is very clear that you feel strongly, but it seems to me that it is a feeling based on emotion, not on a realistic reckoning of the facts.' Washington was trying to get a military advantage, he said. The purpose was not only to get a first-strike capability, but also to place offensive weapons in space:

"I've explained to you that when we had a monopoly on nuclear weapons, we didn't use it", replied the President. "Why don't you trust me?"

"Do you believe that I wouldn't attack you?", asked the General Secretary.

"Well, I've got quite a bit of reassurance on that during this session", said the President. "But any American leader has to base his plans upon the other side's capabilities."

The General Secretary restated his position, repeating charges that Washington was bent on damaging the Soviet economy and that its real interest was to achieve military superiority in space. "You can see I am very excited about this", he said.

There was another silence. The General Secretary eventually said that it appeared an impasse had been reached. The President urged him to reconsider his offer to share research and technology and to open the laboratories. To break the tension, he suggested they go outside for a stroll.

The two men got up and left the fireside. After a while, the President pointed out that at least they had reached agreement on some points.

"Yes," replied the General Secretary.

"Wouldn't further talks be useful?" asked the President. "I would like to invite you to meet in Washington."

"And I would like to invite you to meet in Moscow."

"I accept."

"I, too, accept."

*   *   *

It was hot, bright and sunny. Susan Irvine, a housewife and a mother of three, was on holiday in Geneva. At the same time, the UN

Committee on Disarmament, the only multilateral arms negotiating forum, was in its summer session. The negotiating chamber was fifteen minutes walk from where she was staying. Susan Irvine decided the time had come for her to intervene personally in the long drawn-out talks.

She decided to seek appointments with the British, American and Soviet Ambassadors. She also wrote to French and Chinese missions. After she had sent the letters, she began to feel apprehensive: how could a mother of three, without professional qualifications or title, lay claim to the time of ambassadors? She consulted a friend, who reassured her with the thought that 'people are essentially 65 per cent water. Ambassadors are too'. Duly heartened, she began to make telephone calls. She found it easy to obtain an interview with the British Ambassador. By persistence, she eventually persuaded the American Ambassador to meet her. But all her calls to the Soviet Ambassador drew a polite refusal. So she waylaid him personally outside the UN building. He then agreed.

She found the three Ambassadors were very different men. The Briton was an urbane and polished diplomat of the old school, charming but evasive. The American was tough, direct and hard-line. The Russian was bluntly honest, expressing a total lack of hope in the achievements of the Committee, while emphasising the desperate urgency of military disarmament. Yet she recieved almost the same replies from all three to her prepared questions. Each one declared that 'the other side' was well ahead in the arms race. Each emphasised that 'their side' had taken unilateral disarmament measures in the past, only to have them ignored or taken advantage of. Each considered they were getting a bad press from the world's media.

She concluded: 'It was apparent that East and West were hopelessly deadlocked, each reproaching the other with the same faults, that Britain held hands with the USA and that the five nuclear powers were at odds with the overwhelming wishes of the majority of the countries at the UN.'

*   *   *

The Geneva summit brought together three different strands in the arms control talks which the United States and the Soviet Union had been pursuing together, bilaterally, since 1969. These were the Intermediate Nuclear Forces (INF) talks, the Strategic Arms Reduction Talks (START) and talks on space weapons.

The INF talks had revolved around the Cruise, Pershing II and SS-20 missiles, though occasionally they ranged on to other systems and on to British and French weapons. The START talks concerned 'central strategic systems' – the long-range missiles and bombers of each side. The space talks concerned anti-satellite weapons and the new weapons systems being developed in the Strategic Defense Initiative and in Soviet space programmes.

In the past, a number of arms control and disarmament treaties had been signed. The Partial Test Ban Treaty of 1963 had banned atmospheric testing. The SALT I agreement of 1972 had set limits to US and Soviet strategic weapons launchers. The ABM Treaty of 1972 had limited the development of anti-ballistic missile systems. And the SALT II treaty of 1979 had introduced equal ceilings on the number of nuclear weapons launchers on each side. SALT II had not been ratified by the United States, but nevertheless it was substantially observed by both superpowers.

None of these agreements had reduced the number of weapons in either of the superpowers' arsenals. Nor had they done anything to stop the development of qualitatively new types of weapons. Nevertheless, by setting limits to a potential all-out race, they are important. It is a positive factor that the superpowers are capable of negotiating and reaching any agreements at all in this area.

Recently the superpowers have begun to negotiate over actual reductions in weapons deployments. Progress in these talks has been slow and halting, not surprisingly, since there are many difficulties in the way of an agreement.

One difficulty is that the weapons of one side cannot be directly compared with the other. Each arsenal contains a different mixture of weapons systems, and the weapons systems have different characteristics. This has made it difficult to agree even on units of measurement. Another difficulty lies in the geographical and geopolitical differences between the Soviet Union and the United States. The Soviet Union regards weapons based in Western Europe as equally threatening as weapons based in the United States, since both can reach its territory, whereas the United States regards intermediate-range weapons as 'theatre' weapons and only intercontinental-range weapons as strategic. A third difficulty lies in the relations between the United States and the Soviet Union, which are at times so tense as to make negotiations impossible. A fourth lies in different perceptions of the role of the nuclear forces of the other three nuclear weapons states, France, Britain, and China. The Soviet Union has attempted to include the British and French forces in an overall agreement with the United States, while the United States has insisted on excluding them. France, Britain and China themselves have refused to enter the negotiations. Added to all these difficulties have been internal disagreements and scepticism within some nuclear states as to the value of negotiating arms control and disarmament agreements at all.

Nevertheless, the negotiating teams have persisted, and positions have developed. Before the Geneva talks, the position of the United States was that each superpower should make deep cuts in its strategic arsenal, with a limit of 5000 nuclear warheads on each side (from a starting point in 1986 of about 10 000 each). The United States wanted a complete global ban on intermediate-range nuclear weapons, although various permutations of this position had been put forward in the INF talks. On space weapons, the United States was not prepared to abandon its Strategic Defense Initiative.

The Soviet position, introduced in a dramatic new initiative launched by General Secretary Gorbachev shortly before the talks in Geneva began, proposed a cut of 50 per cent in nuclear weapons capable of reaching each other's territory, a ban on space weapons, and a separate deal on intermediate nuclear forces. Later, in January 1986, Gorbachev proposed a three-stage plan for complete multi-lateral nuclear disarmament by the year 2000. The Soviet Union would accept a ban on intermediate nuclear forces in Western Europe so long as the French and British nuclear forces were not modernised. To emphasise his seriousness, Gorbachev maintained a unilateral moratorium on testing while the response to his offer was being considered.

The United States continued nuclear testing and pressed ahead with its Strategic Defense Initiative, while exploring the possibility of a separate deal on intermediate-range weapons. The French and British declared that their independent nuclear weapons were not negotiable in the INF talks. Meanwhile the Soviet Union was not prepared to abandon its SS-20s in the East, facing China, which had not yet been a party to the exchanges.

It was evident that the nuclear arsenals and negotiating positions of all five nuclear weapons states were related in these discussions. Despite the sharp disagreements, there seemed to be some basis in 1985 for a real multilateral disarmament agreement, based on the principle of 50 per cent cuts in the superpowers' arsenals, which could then lead to the involvement of the other three major nuclear weapons states in the arms talks. But in order to take advantage of this opportunity, all the parties would have to have a serious desire to reach an agreement. It seemed that, despite protestations in favour of arms control by the governments of the five nuclear weapons states, the prospects for reaching such an agreement were highly uncertain. The way in which arms control and disarmament positions are formulated within the five governments may help to explain why this is so.

*   *   *

Once or twice a week, the men who make the crucial decisions on US arms control positions meet in the windowless situation room of the White House basement. The group is known as SAC-G, the Special Arms Control Group. The chairman is the National Security Adviser. In 1985 this was Robert MacFarlane, a former lieutenant-colonel in the Marines. From the Pentagon, the Under Secretary for Policy, then Fred Ikle, and the Assistant Secretary for International Security Policy, then Richard Perle, attend. The Joint Chiefs of Staff have a seat, and officials from the State Department and the Arms Control and Disarmament Agency attend. Paul Nitze had a seat too in 1985, and when they were in Washington, members of the US delegation to the Geneva talks would sit in on the meetings.

It is at these inter-agency meetings that US positions are hammered out. Although the members of SAC-G are theoretically equals in shaping US policy, in practice some members carry greater weight than others. Richard Perle, the clever and assertive Assistant Secretary of Defense, was the most effective member of the group, being always well-informed, self-confident and highly articulate. He held earlier American arms control efforts in disdain and advocated the most hard-line of agreements: anything less, he argued, would compromise US security. The positions he took were, more often than others', the positions which eventually became the US negotiating offers. Some US officials have expressed the opinion that a break-through in arms control will be impossible while Perle remains in a position of influence.

Perle's view of the SALT talks was that they had permitted the Soviets to gain an advantage in heavy missiles, and the only kind of agreement he was interested in was one which would reduce these missiles. 'We've had several agreements and treaties', he said, 'and I think it's fair to say in the main that they have failed to accomplish (their) purpose because the threat has grown, and grown significantly. The Soviets now have some eight thousand strategic nuclear warheads that they did not possess when the talks began, in November of 1969, that led to the SALT I agreement. Four thousand of those warheads have been added since '79 when the SALT II agreement was reached, and that is hardly a successful result for arms control, even though agreements had been signed . . . If the agreements diminish the threat, diminish the burden that we must bear in maintaining the stable balance, then they're highly advantageous, but we haven't yet achieved an agreement of that character. I think the Soviets see arms control negotiations principally as a political device to permit their strategic and conventional force build-up to continue. That's the only conclusion one can arrive at by studying the numbers in parallel with the history of the negotiations.'

The State Department, although it is responsible for foreign policy, has been less influential than the Pentagon. Within the State Department, the Office of Politico-Military Affairs has been the key department for preparing arms control. Its responsibility is to co-ordinate diplomacy with defence. Its director, until 1985, was Richard Burt, a defence intellectual who had worked on the *New York Times* and at the International Institute for Strategic Studies in London before being picked by Haig, Reagan's first Secretary of State. Burt was not so opposed to SALT II as Perle and he wanted the US to arrive at a position that was negotiable with the Soviet Union. He suggested that the United States should trade its modern MX missile for the heavy Soviet SS-18s. The proposal made no headway in the Administration. The Defense Department, which had been fighting to get the MX deployed against opposition from the Congress and the public for many years, was not prepared to make its newest missile a hostage to arms control.

The Arms Control and Disarmament Agency played a relatively minor role. When the Reagan Administration came into power, it had filled the Agency's top posts with members of the Committee on the Present Danger, a pressure group opposed to SALT II and arms control. The Agency's computer had been removed and its research library was given away to a university. The Agency's officials, who previously had usually followed the State Department's line, now generally followed the Pentagon's.

The Joint Chiefs of Staff represented the voice of the military, which did not always coincide with that of the civilian officials in the Department of Defense. The Joint Chiefs of Staff had supported the SALT agreements. Indeed, their support was crucial to President Carter in pursuing SALT II. Although the Joint Chiefs did not often initiate arms control positions, their views were important. Any treaty on arms control would need a two-thirds majority in the Senate to be carried, and if the professional military testified that a treaty was inimical to US military interests, it would be unlikely to win the Senate's support. At first, the Joint Chiefs supported the State Department proposals over START, since in their view only a combination of cuts in launchers and cuts in warheads on both sides would enable them to cover all the targets they had been allocated in the SIOP targeting plan. They would not, however, support any proposals which threatened the development of existing missiles like the MX. Nothing in SALT II was an obstacle to the rearmament programme, asserted General Jones, the Chairman of the Joint Chiefs of Staff. 'If SALT II disappeared, there's nothing we'd do differently.'

All of the members of the SAC-G were either military officers or political appointees to the bureaucracy. Sometimes, in the event of disagreement, issues would be taken to the National Security Council, or ultimately to the President for a decision. The Office of the Secretary of Defense, the State Department and the other agencies often disagreed, and US policy was made usually by the outcomes of power struggles between them. At times these would be resolved by the President, though in the Reagan Administration his own personal views on these complex issues were often so unclear that others either inserted passages into his speeches representing their own positions or interpreted his general statements in a way which suited them.

Congress, too, sometimes plays an important role. Its Foreign Relations Committees monitor the progress of the Administration's positions on arms control, and when there is a major change of position this is usually followed by a briefing for key members of Congress. Very occasionally, the Congress actually initiates elements in US positions. For example, during the START talks the Congress put forward its own proposal for 'build-down', a scheme to reconcile modernisation and reductions by requiring the removal of two old missiles for every new one. For a time, Congress had a strong bargaining position, since it was about to vote on the development of the MX, which many opposed. A deal was done that Congress would support the MX only if the Administration adopted the Congressional arms control scheme. A variant of the Congressional scheme duly became part of the US position.

The differences over arms control between the various agencies of the government became clear again in January 1986, when Gorbachev put forward his three-stage disarmament proposal. The Pentagon took the view that there was nothing new in the Gorbachev package, and there was no need for the Administration to change its position. The State Department suggested the most flexible response, proposing concessions on both intermediate-range and strategic weapons. The Arms Control and Disarmament Agency put forward a middle position, suggesting that the US respond only to the part of the offer which concerned intermediate-range missiles, on which there seemed hope for an early agreement. This position was eventually accepted after a month of infighting among the agencies.

It has been said that it is harder to get agreement on arms control within the US government than it is between the United States and the Soviet Union. Whether or not this is so, an official position first has to clear the inter-agency process, and thus has to be acceptable to either the Office of the Secretary of Defense or the Joint Chiefs of Staff, or both. Under the Reagan Administration and its predecessors, the Pentagon has been more powerful than the State Department in controlling arms control policy. Consequently, those who are in charge of the development and ordering of nuclear weapons also have a determining say in the preparation of measures which are intended to restrain them. Many of those who have served in the US negotiating teams have been drawn from the 'weapons community'. For example, Richard Wagner, the Associate Director of Livermore and Director of the Nevada Test Site, and now Assistant Secretary of Defense (Atomic Energy), served on the US SALT I planning team. T. K. Jones, the Deputy Under Secretary Research and Engineering (Strategic and Theater Nuclear Forces), who formerly worked on missiles for Boeing, also served on the SALT I team. Paul Nitze, the chief US negotiator at the INF talks in the early 1980s, had previously served on the Gaither Committee, which had recommended a build-up of missile forces in 1957, and on the Committee on the Present Danger, which had recommended the Reagan Administrations's arms build-up. Senator John Tower, who headed the US strategic weapons negotiating team at Geneva, had been the chairman of the Senate Armed Services Committee and a prominant 'hawk' in Congress.

There is no corresponding mechanism in the US government for those who favour arms control to exercise an influence over the development and acquisition of new weapons. The Arms Control and Disarmament Agency was excluded from meetings called to decide the future of the MX and the B-1. Information on new weapons is often withheld from the State Department. The State Department and the Arms Control and Disarmament Agency do not see the Defense

Guidance, the document with which the Secretary of Defense launches the defence budget process, before it is too late to influence it. There is only one formal mechanism through which the government takes arms control considerations into account in weapons decision-making. This is a requirement that the government produce an annual Arms Control Impact Statement to the Congress. This document is produced by the Arms Control and Disarmament Agency and has no impact on weapons decisions.

The Arms Control Impact Statement for Fiscal Year 1986 gives eloquent testimony of the limited impact of arms control on weapons decision-making. On SDI, the report says:

> The Strategic Defense Initiative research programme submitted with the FY 1986 budget is consistent with all US treaty obligations, supports US arms control and national security policy, and is consistent with the US position in current arms control negotiations.

On cruise missiles:

> The sea-launched cruise missile is consistent with US arms control policy and related Presidential decisions and is not constrained by existing arms control agreements'.

On air-launched cruise missiles and the B-1 bomber:

> Programs to ensure the continued effectiveness of the airborne strategic offensive force support US arms control and national security goals of deterring war and preventing coercion while maintaining international stability.

There is a community in favour of arms control in the United States, but this community has not been in charge of arms control policy. Under the Reagan Administration, the arms control advocates were edged out of the posts that mattered for formulating the government's position by appointees opposed to arms control. Even in previous administrations, the arms controllers had not been able to assert their influence if the military and civilian officials in the Pentagon opposed them. While the arms control decision-making arrangements within the US Government remain as they are, it seems unlikely that arms control will act as a serious restraint to the development of new nuclear weapons in the United States.

*   *   *

The situation in the world today is highly complex, very tense. I would even go so far as to say it is explosive.
Certain people in the US are driving nails into the structure of our relationship, then cutting off the heads. So the Soviets must use their teeth to pull them out.

With Gorbachev's accession to power, the Politburo made an orchestrated change in the direction and vigour of its arms control efforts. Arms control policy, as well as nuclear weapons policies, are decided at the top, unlike in the United States. 'Our decision-making system differs from the American in that it is more centralised', said Valentin Falin, a member of the Central Committee Secretariat. 'In international or national security affairs the American Secretaries of State and Defense can make a good many decisions on their own. In our case all foreign policy and national security questions must be discussed and decided in the Politburo.'
Beneath the Politburo, the Defence Council plays a role in the formulation of arms control policy, and it is thought that agreements are sometimes worked out between the political leadership and the military in this forum. There is also believed to be a subgroup of the Politburo which specialises in arms control issues. The major policy decisions are taken between these three groups – the Defence Council, the Politburo arms control group and the full Politburo. By this means, the top political leadership retains close control over arms control policy formulation, and this control probably extends to the consideration of detailed proposals for the Geneva talks. Staff work is carried out by the Secretariat of the General Secretary and by several departments in the thousand-strong Central Committee Secretariat. There the International Affairs Department, the International Information Department and the Defence Industry Department are likely to be involved. Of these, the last has authority to request information from the General Staff, the military section of the central planning agency GOSPLAN, the defence industry ministries and the Deputy Defence Ministers, who are the military Chiefs of Staff of the five Services. Such information would be necessary to assess current Soviet weapons production and plans, and so the Defence Industry Department may have a key role. If this is so, the same decision-makers who are responsible for the production of nuclear weapons systems also

have a pivotal place in the system for formulating arms control.

Besides these Party bodies, the most important source of advice on arms control is the General Staff. The General Staff has much more authority over the formulation of Soviet arms control positions than the Joint Chiefs of Staff have over American positions, and hold a monopoly of professional military expertise and military judgement. Details of Soviet weapons systems are so closely guarded that even the Soviet diplomats at the SALT talks were not fully informed about Soviet military deployments. On one occasion, a Soviet general – Colonel-General Ogarkov, who later became the Chief of the General Staff, asked the US delegation not to discuss certain details of Soviet weapons in a Soviet diplomat's presence. 'Such information is strictly the affair of the military', said Ogarkov. Officers from the General Staff were members of the Soviet delegation in both the SALT talks. The military were relied on for detailed advice even at the highest levels, as this comment by Kissinger on the SALT II talks indicates: 'The meeting started with two generals sitting behind Brezhnev. Whenever we started getting concrete they started slipping little pieces of paper to him, or butting into the conversation one way or the other. So, at the first break, Dobrynin came to me and asked whether we couldn't confine the meeting to three people on each side – which got rid of the two generals. So after that, whenever numbers came up, we would explore the numbers, and then he would take about a forty-five minutes break, either to consult the two generals or, on at least two occasions, to consult Moscow.'

Two other sources of military advice are important. One is military intelligence, carried out by the Main Intelligence Directorate (GRU), probably with contributions from the KGB. This agency would serve as a source of information about American military programmes and research plans, and its interpretations could be expected to influence Soviet thinking. Clearly, for example, the Soviet Union regards the purpose of the Strategic Defense Initiative to be the development of offensive 'space strike' weapons rather than the stated purpose of a defensive shield, and this analysis could be consistent with a 'worst-case' military assessment of US capabilities. The other source is the Ministry of Defence, which probably has a smaller role in arms control policy than the General Staff, but appears to be consulted at times by the negotiating teams, perhaps on procurement matters.

The Ministry of Foreign Affairs provides the diplomats who head the Soviet delegations. Its role is to implement rather than to formulate policy, and the Ministry of Foreign Affairs is not given the full information which would enable it fully to participate in the policy process. It may, however, contribute general political and diplomatic advice. In addition, the Soviet leadership has been placing increasing reliance on civilian arms control experts in bodies like the Institute of the USA and Canada and the Institute of World Economy and International Relations. Both these institutes keep in touch with international opinion on arms control issues, and they may thus serve as 'ears' to the Politburo, as well as alternative sources of advice to the military's. These civilian advisers are better placed to judge the likely Western response to Soviet offers than professional military men, and the directors of these Institutes, such as Georgii Arbatov, sometimes have close links with key members of the Politburo.

The Soviet negotiators themselves are career diplomats. Though they are skilled negotiators, they are given less freedom of manoeuvre at the talks than their American counterparts. The head of the Soviet START delegation, Viktor Karpov, was a veteran of both the SALT and the INF talks, and an experienced and accomplished negotiator. He enjoyed drinking, and sometimes appeared inebriated at the talks, once almost falling off his chair in a plenary session.

Compared with the US system, Soviet decision-making on arms control is coherent and centralised. Instead of policy being made in a number of separate agencies, who then struggle to establish their own position as government policy, Soviet policy is co-ordinated and formulated in the Politburo, the Defence Council and the Central Committee Secretariat. In the Soviet Union the political leadership is in charge, whereas in the United States the Pentagon is the most important influence on decisions. Partly for this reason, arms control policy is broader and more political in the Soviet Union, narrower and more technical in the United States. In the Soviet Union, as in the United States, it seems unlikely that the government would take a position that had strong opposition from the military, and there is considerable overlap between those who are responsible for decisions on nuclear weapons programmes and those who formulate arms control positions. But, because in the Soviet Union these policy issues are dealt with at the highest level, there is a forum in which the political authorities can compare the contribution made to security by new weapons programmes with those which could be made by arms control. In the United States, because the arms control process does not touch weapons acquisition, no such forum exists.

*   *   *

'We really mean it when we say that we want to negotiate arms control agreements', Mrs Thatcher had said, on the day the Cruise missiles arrived. 'Britain is ready to pursue, in the right circumstances, a dialogue with the Soviet Union and the countries of Eastern Europe.'

'Multilateral disarmament and arms control form an essential element of our foreign policy', she said on another occasion. 'Responsibility therefore rests with . . . the Secretary for Foreign and Commonwealth Affairs, in consultation with . . . the Secretary of State for Defence. Within the Foreign and Commonwealth Office the Minister of State has special responsibility for arms control and disarmament.'

The same year, Lord Carrington, then the Foreign Secretary, also made a statement about arms control. 'Arms control is part of our national security policy', he said. 'Our arms control endeavours and our defence effort are two different but complementary ways of achieving the same end: peace with freedom to pursue our legitimate national interests around the world.'

In Britain two government departments are responsible for arms control policy, the Foreign Office and the Ministry of Defence. The Foreign Office is nominated the 'lead' department and a junior minister at the Foreign Office has responsibility for the work. At times Foreign Secretaries have taken a vigorous interest in arms control – David Owen, for example, was active in this area, but most Foreign Secretaries do not involve themselves in the matters of detail.

Inside the Foreign Office the two concerned departments are the Arms Control and Disarmament Department and the Defence Department. The latter is responsible for 'liaison with the Ministry of Defence on international aspects of defence', though in practice both departments liaise with MoD on arms control. The Defence Department gained importance in 1979 when the NATO section of the West European Department of the Foreign Office was transferred to it, so the British attitude to the INF and START talks fell into its remit, as did the Mutual Balanced Forces Reduction talks and the Confidence Building Measures discussed in the Conference on Security and Co-operation in Europe. The Arms Control and Disarmament Department (ACDD) has responsibility for the British presence at the international Conference on Disarmament at Geneva and at various UN fora. Beneath the ACDD is an Arms Control and Disarmament Research Unit, once a prestigious body, now reduced to two researchers and a budget of £80 000.

Both the Arms Control and Disarmament Department and the

Defence Department are responsible to a Superintending Secretary, who reports to the Permanent Undersecretary and to the Minister of State. One Superintending Secretary once wrote on a paper, 'I know only one thing about disarmament, and that is, that it is not a UK interest.' The official was later relieved of this responsibility.

All arms control policies developed within the Foreign and Commonwealth Office have to be cleared within the Ministry of Defence. Until 1985, papers used to go to Defence Secretariat 17, a body responsible for 'national and NATO nuclear policy'. Following a re-organisation in 1985, its arms control functions were transferred to a newly-created Defence Arms Control Unit. In 1981 a junior Defence Minister stated: 'Excluding clerical and support staff, about 40 people are employed by the Ministry of Defence wholly or substantially on arms control and disarmament.' This figure includes seismologists working on problems of detection, for a test ban, at Blacknest in Berkshire and toxicologists working at Porton Down.

Within the Foreign Office, the Defence Department has more influence than the Arms Control and Disarmament Unit. Moreover, since the Ministry of Defence clears all Foreign Office documents on disarmament, the Ministry of Defence effectively has a veto. Senior officials in the Ministry of Defence, such as the Assistant Chief of Defence Staff (Programmes), provide technical and scientific advice on arms control positions and the approval of the Chief of Defence Staff and the Permanent Undersecretary is required to 'clear' the Foreign Office papers.

There is further co-ordinating machinery at the Cabinet Office level, where the Defence and Overseas Policy Committee of the Cabinet, chaired by the Prime Minister, is responsible for arms control and disarmament. It is 'shadowed' by a committee of civil servants, the Overseas Policy and Defence (Official) Committee, which has a subcommittee on Disarmament and Arms Control. The Permanent Undersecretary of the Ministry of Defence and the Superintending Undersecretary of the arms control department within the Foreign Office liaise at this level.

Just as arms control positions have to be cleared with the MoD, so British positions have to be cleared with NATO. The NATO allies are consulted on policy changes and initiatives, not only at ministerial level but also at the level of working groups inside the Defence and Foreign ministries.

The Prime Minister personally has a great deal of influence over the formulation of policy. Offers and responses are sometimes communicated directly by correspondence between Heads of State. The personal views of the Prime Minister are therefore influential. According to the US Ambassador, Kingman Brewster, Mrs Thatcher's attitude was 'a grudging acceptance that some sort of arms control is a necessary concomitant (to deploying Cruise and Pershing missiles) but an insistence that it not hold up modernisation'. Mrs Thatcher has been critical of the Foreign Office and was dissatisfied with a draft of a speech Foreign Office officials prepared for her for the UN Second Session on Disarmament. Instead, she turned to Michael Quinlan, the Deputy Under Secretary at the MoD, who had piloted through the Trident decision on her behalf.

Since 1979 Britain has sent three papers to the Conference on Disarmament in Geneva, relating to the Comprehensive Test Ban Treaty. No other official initiatives in the field of nuclear arms control have been made in any international forum.

An opportunity to make an initiative arose in 1986, when Mr Gorbachev proposed direct negotiations with Britain and France. Mrs Thatcher replied that Britain was not prepared to enter direct talks with the Soviet Union at that stage. In a comment, the Foreign Secretary, Sir Geoffrey Howe, said: 'If the Soviet Union's and the United States' strategic arsenals were to be substantially reduced, and if no significant change had occurred in the Soviet Union's defence capacity, Britain would want to review her position, and see how she could best contribute to arms control.'

Later, in January 1986, Mr Gorbachev made his proposal for a three-stage reduction in nuclear arms. Mrs Thatcher replied that Mr Gorbachev's proposals had been considered with great care and discussed with Britain's allies. She said that to simply describe the goal of a nuclear-free world and attach a timetable to it was not a practical approach. For the forseeable future nuclear weapons would continue to make an essential contribution to peace, she said. She rejected a Soviet proposal that Britain should refrain from modernising its nuclear forces in exchange for cuts in Soviet intermediate-range missiles. In the meantime, she expressed support for the American position on strategic and intermediate-range weapons.

The British reluctance to negotiate represents an odd change of position. Britain acquired its nuclear weapons in order to secure a seat at the negotiating table. Now that the arms negotiations appear to threaten British nuclear weapons, the government is unwilling to take a seat at the talks.

*   *   *

'If there should one day be a meeting of States that truly want to organise disarmament – and such a meeting should, in our mind, be composed of the four atomic powers – France would participate in it wholeheartedly. Until such time, she does not see the need for taking part in proceedings whose inevitable outcome is . . . disillusion.' With these words, General Charles de Gaulle announced the French withdrawal from the Eighteen Nations Committee on Disarmament in 1962 and inaugurated the 'empty chair' policy of France on arms control negotiations. France refused to sign the Partial Test Ban Treaty and continued to carry out atmospheric tests until 1974.

In its odd role as a *de facto* member of the Western Alliance and yet a non-member of NATO, France has been able to continue to maintain its 'empty chair'. Since it was not a party to the agreement to modernise NATO forces, France has not been involved in the INF talks, and President Mitterand rebuffed General Secretary Gorbachev's proposal for bilateral talks as firmly as Mrs Thatcher had done. 'It would not be reasonable to expect a negotiation', said Mitterand. The French defensive forces were kept at a minimum, he said, and there was 'no margin for manoeuvre'. France would contemplate cuts in its arsenal only if the Soviet Union first made sharp cuts in Soviet nuclear weapons and refrained from deploying strategic defences. However, while ruling out negotiations, Mitterand said he would 'not exclude a dialogue'.

The President himself is probably the most important figure in framing French policy on arms control and disarmament. Within the government, there are arms control units in the Foreign Ministry and at the Secrétariat Général de la Défense Nationale (SGDN), an advisory body dominated by the military which is attached to the office of the Prime Minister. The influence of the Foreign Office is circumscribed by the defence and military interests. As in Britain, the organisations concerned with arms control and disarmament have a low profile in the government.

The French Parlement has no role in formulating arms control policy. No parliamentary hearings or inquiries are held on this subject.

*   *   *

As everyone knows, the two big nuclear powers possess more than 95% of the world's nuclear arsenal, posing a serious threat to humanity. It is they who should take the lead in tangible actions in nuclear disarmament. Only under such circumstances would China take part in a widely representative international conference with all nuclear countries, discussing the practical moves of common nuclear disarmament up to the complete prohibition and thorough destruction of nuclear weapons.

This statement, by Li Xinnian, the President of the Chinese People's Republic, puts the Chinese position on arms control and disarmament. The Minister of Foreign Affairs in 1982, Huang Hua, who chaired the Chinese delegation to the Special Session on Disarmament of the UN General Assembly, said that 'if the two superpowers take the lead in halting the testing, improvement or manufacture of nuclear weapons and in reducing their nuclear weapons by 50 per cent, the Chinese government is ready to join all other nuclear states in undertaking to stop the development and production of nuclear weapons and to further reduce and ultimately destroy them altogether'.

Since most of the medium-range nuclear weapons of the Chinese forces are directed against a potential attack on China by Soviet forces, an improvement in Sino-Soviet relations is critical to real progress in arms control for China. China has imposed three conditions for such an improvement: the Soviet Union should first withdraw its million troops along the Chinese border, get out of Afghanistan, and stop supporting Vietnam in Kampuchea.

Although China has not been involved in the arms talks, which are seen as being dominated by the superpowers, China is increasing its participation in peace and disarmament meetings. A Chinese People's Association for Peace and Disarmament has been set up, to improve understanding and support within China for disarmament and to foster links with peace movements in other countries. 'Our one billion Chinese people are prepared to join hands with peace movements the world over in unremitting efforts for world peace', said General Secretary Hu Yaobang. The Chinese no longer consider that nuclear war is inevitable – so long as the peoples of the world vigorously resist the arms race. The Chinese leaders want a period of peace, which China needs in order to modernise. Meanwhile, however, the Chinese nuclear force is being modernised too – to oppose the hegemony of the superpowers.

*   *   *

A rare moment of opportunity arose in the arms control and disarmament discussions in 1985, when the United States and the

Soviet Union agreed to the principle of negotiating 50 per cent cuts in their strategic arsenals and the Soviet Union halted nuclear testing. The positions of both sides remain far apart, but the 50 per cent principle is important as it offers a basis for involving the other three powers in the negotiations.

If there is to be progress in disarmament, all five nuclear states will have to be involved. Their positions are related, by the long range of their weapons. For example, the question of the Soviet SS-20 missiles affects Britain and France and China. The development of new space weapons is obviously a matter of concern to the smaller nuclear powers as well as the superpowers.

There is little prospect that the opportunities that now exist for progress in arms control and disarmament will be taken while those who are responsible for developing new nuclear weapons exercise, in effect, a veto over their governments' formulation of arms control and disarmament positions. Only if there is greater public involvement and pressure for real disarmament agreements is progress likely to be made. A useful starting point would be to examine and overhaul the institutional arrangements for arms control and disarmament.

Susan Irvine will be one of those demanding more action on disarmament from her government. 'If old avenues have failed, we must demand that our representatives try new ones', she says. 'Innovation and imagination need not be the prerogatives of warfare and defence only . . . three years ago I used to sigh fatalistically, "Well, if it has to be, it will be. There's nothing I can do about it." Now I say, "Well, if it happens, it will be despite me!" and I can look myself and my children in the eye.'

12 Conclusions

[The Buddhists outside NATO – power of the nuclear weapons lobbies – differences in decision-making between East and West – a window of opportunity for disarmament – proposals for change – envoi]

The banners fluttered gaily, and the wind ruffled the saffron robes of the Japanese Buddhist monks. They walked slowly together, in twos, holding the flat circular drums in their left hands and short curved sticks in their right. With their shaven heads, and their faces half smiling, half sad, they seemed timeless, without age. They banged their drums constantly in a simple rhythm, chanting at the same time: 'Namu Myo Ho Ren Ge Kyo'.

Inside NATO headquarters, the diplomatic and military aides were clustered around a long, wide conference table. They wore ties and suits, and many held papers in their hands. There was a subdued murmur of conversation. They were talking in groups of twos and threes, perhaps about Euromissiles, or battlefield nuclear weapons, or neutron bombs. Some stood with arms folded, others clenched their hands behind their backs, others were making vigorous hand gestures to emphasise a point.

If they heard the Buddhists banging their drums outside, they ignored them. And the more they were ignored, the louder the Buddhists banged.

*   *   *

This book has described how defence ministry officials, military leaders, defence industry managers and nuclear scientists take decisions, independently and jointly, in a maze of committees whose deliberations are secret and whose names are little known. Together with top level politicians, these people shape policies and decisions which have momentous consequences for peace and war. Yet they are subject to little real accountability. Selected from communities with shared values, insulated from the public by their official status and by secrecy, they have been able to ignore hitherto the growing swell of dissent. The potential restraints, through parliamentary bodies, and informed public opinion, financial scrutiny and arms control, have yet to prove as strong as the forces driving the arms race. However, as nuclear policies arouse growing controversy, the balance of these forces may change.

This book has shown that there are striking differences in the ways in which the nuclear circles govern, especially between East and West. In the West the tendency has been for power to accrue to the permanent bureaucracies, the nuclear laboratories and the military services, which have the institutional continuity to see projects through from beginning to end. The politicians come and go rapidly and lack the power to make decisions stick if the bureaucracies oppose them. In the East, in contrast, the continuity of the Politburos and Central Committees has given control over nuclear weapons programmes to the political leadership.

A second fundamental difference lies in the level of political debate. Whereas in the United States and Britain, and to a lesser extent in France, there is political controversy about nuclear weapons pro-grammes and the public has the power to bring about change through the electoral system, in both China and the Soviet Union the expression of public opinion is subject to the sanctions of Party discipline and there are no direct means whereby public opinion can alter policy.

In the West, the political leadership is accountable to the public, but the control of the political leadership over nuclear weapons is weak. In the East, the control of the political leadership over nuclear weapons is strong, but the political leadership is not accountable. In none of the five societies are decisions about nuclear weapons accountable to the public.

The people and organisations who form the inner core of nuclear decision-making, as this book shows, have strong predispositions to develop nuclear weapons. There are, however, powerful bodies which could exercise restraint. In the West, there are parliaments and Congresses, departments inside the government but outside the nuclear area, the critical press, the political parties, the unions, the churches and the many other non-governmental organisations through which public opinion can be voiced. In the East, there are organs of the Party and the State outside the defence industry, Institutes and foreign policy bodies, and non-governmental organisations. Outside the nuclear weapons states there are governments and peoples of the non-nuclear countries. If the pressures for change become sufficiently strong, the lobbies for nuclear weapons can be brought under control.

In the East, the organisations which are responsible for producing nuclear weapons, though powerful, are neither so strong nor so independent as they are in the West. They can be controlled by the Politburo. In the West, the weapons lobbies, though powerful, are neither numerous, nor independent, nor autonomous. Their apparent strength reflects the weakness of the restraints on them more than their intrinsic power. Indeed, the weapons lobbies comprise elements with different interests which frequently conflict. The rationales they use to justify new weapons are often inconsistent with one another. If the secrecy which protected them were stripped away and the rationales for their actions searchingly probed and scrutinised, it seems unlikely that the weapons lobbies would continue to act as they now do.

Two bodies appear to have sufficient freedom of action to impose restraint. In the West, public opinion, through parliamentary institutions and elections, has the power to control the bodies which develop nuclear weapons. In the East, the Politburo and the leading organs of the Party have such power now.

*   *   *

Some time before he was murdered on the streets of Stockholm, Olaf Palme said, 'It is very unlikely that disarmament will ever take place if it must wait for the initiatives of governments and experts. It will only come about as the expression of the political will of people in many parts of the world.'

An examination of the process whereby arms control and disarmament positions are formulated by the governments of the nuclear states suggests why the first part of Mr Palme's statement is likely to be correct. In the United States, Britain and France, the defence ministries effectively have a veto over national arms control positions. In the Soviet Union, the General Staff has considerable influence too, though it appears that the Politburo is in overall control.

Periodically in the history of the arms race, there has opened a 'window of opportunity' for achieving restraint. The first was in 1945–46, when international control of nuclear weapons was considered and the UN General Assembly unanimously resolved to eliminate atomic weapons from national armaments. Another occurred in 1955, when a plan for nuclear disarmament seemed briefly to be acceptable to both the United States and the Soviet Union. There is another opportunity now. A bilateral agreement between the superpowers is within reach, based on reductions by 50 per cent in the number of strategic weapons, the elimination of intermediate-range missiles, and a moratorium on nuclear testing. Agreement along these lines could open the way to multilateral disarmament measures, of which the most urgent is a comprehensive nuclear test ban.

There has been considerable support for the two ideas of a Freeze and a ban on nuclear tests. Major opposition parties in the United States and Britain support these two measures, and the Soviet Union has also expressed support. There must be some prospect therefore that measures such as these could be introduced.

However, there is strong opposition to these proposals, notably from within the nuclear weapons laboratories, the defence contractors and the defence ministries. Efforts to control the international competition in nuclear weapons must therefore proceed step by step with action to bring the internal weapons lobbies under restraint.

If we are to fashion reins, to restrain the horse on which we are galloping out of control, what must these be?

First, information. Unless the public knows what weapons are being developed, there can be no accountability. The public has a right to know how its money is spent, especially in matters affecting peace and war. The limitations on this right set by official secrecy must be rolled back. Less secrecy could also have a stabilising effect on the arms race, since exagerrated threat assessments have rested in part on ignorance and uncertainty. The right of inquiry should be extended into the nuclear weapons laboratories themselves, in both West and East.

Second, more effective financial control. Even in the United States, where Congress has significant financial powers, it is too easy for the military, the defence contractors and the weapons laboratories to get the resources they want. In Britain, the lack of detailed financial control over nuclear expenditure is urgently in need of reform.

Third, greater separation of responsibilities. Measures should be taken to break up certain powers the weapons community have acquired which are either corrupt or corrupting. The bribery and lobbying of the 'Beltway Bandits' is the most overt feature of this phenomenon; it should be curtailed by law. Representatives of the nuclear laboratories and the defence contractors should not be permitted to participate in policy decisions about nuclear weapons. In a democracy, it is dangerous to allow the military to take decisions about the procurement of its own weapons. It is unhealthy to appoint nuclear weapons scientists and policy-makers as their countries' representatives at negotiations on test ban treaties.

Fourth, governments in the West need to assert their control over the defence ministries. In Britain, for example, it is open to question whether a government elected on a platform of unilateral nuclear disarmament could carry out such a policy in the Ministry of Defence against opposition and NATO. One means of breaking up bureaucratic power is to increase the rotation of officials, and perhaps the Ministry of Defence could be made less insular if its staff were rotated with other ministries. Appointments to top positions could be made from outside the Ministry, as they are in the United States.

Fifth, there is also a need for greater political control of the nuclear weapons laboratories. Their power to initiate completely new weapons systems without political direction is a key destabilising factor. Recently a ban on new weapons research and development has been seriously aired for the first time, in the context of the Strategic Defense Initiative. Such a ban could not only unlock the door to a multilateral arms reduction agreement, it also offers a means to introduce control over the engines of the arms race.

Sixth, there is a need for greater control by parliamentary bodies over the weapons acquisition process. In Britain, the powers of the Defence Select Committee should be increased, and there should be detailed scrutiny and financial control over programmes before the money has been spent. Parliament must be free to inquire about new programmes while they are in their early stages, and its right of scrutiny must extend to the nuclear weapons laboratories and to NATO.

Seventh, the status of arms control and disarmament within the Western governments should be upgraded, particularly in Britain and France. These responsibilities should be firmly in the hands of Foreign Ministries. New weapons should be required to be appraised for their impact on existing and potential arms control agreements, rather than vice versa.

These measures could begin to introduce a greater degree of democratic control into decisions about nuclear weapons. Of course, they would depend on the support of an active constituency of public opinion. The growth of public concern in the Western nuclear countries is a condition of restraint.

*   *   *

In the summer of 1985, a gathering of nuclear physicists, technicians and engineers met at Los Alamos to celebrate the fortieth anniversary of the dropping of the bomb they had made on Hiroshima. After a good-natured party, in which old acquaintances were refreshed, stories told and memories recalled, a speech was made. Atomic weapons, said the old veteran, had finished the war off quickly. 'We won the war', he called. The audience cheered. Then an accordion player struck up a tune and they began to sing:

> You are my sunshine,
> My only sunshine,
> You make me happy when skies are grey.
> You'll never know, dear,
> How much I love you,
> Please don't take my sunshine away.

In 1945 Robert Oppenheimer, who had led these physicists, observed the first atomic explosion, test named Trinity, at Alamogordo. Its flash was 'brighter than a thousand suns'. When the war was over, Oppenheimer campaigned hard to bring the atomic bomb under international control, and to prevent the development of the hydrogen bomb. He observed the early growth of the nuclear lobbies and sensed their power. In 1965, shortly before his death, Oppenheimer was asked what he thought of Senator Robert Kennedy's proposal that President Johnson initiate talks with a view to halting the spread of nuclear weapons. He replied, 'It's twenty years too late. It should have been done the day after Trinity.'

In 1945 there were three nuclear weapons. Now there are over 50 000. It is not open to us to know whether it is too late. But at any time, we have the power to see, think and act.

In the nuclear age, every day is the day after Trinity.

Index